Springer Theses

Recognizing Outstanding Ph.D. Research

For further volumes:
http://www.springer.com/series/8790

Aims and Scope

The series "Springer Theses" brings together a selection of the very best Ph.D. theses from around the world and across the physical sciences. Nominated and endorsed by two recognized specialists, each published volume has been selected for its scientific excellence and the high impact of its contents for the pertinent field of research. For greater accessibility to non-specialists, the published versions include an extended introduction, as well as a foreword by the student's supervisor explaining the special relevance of the work for the field. As a whole, the series will provide a valuable resource both for newcomers to the research fields described, and for other scientists seeking detailed background information on special questions. Finally, it provides an accredited documentation of the valuable contributions made by today's younger generation of scientists.

Theses are accepted into the series by invited nomination only and must fulfill all of the following criteria

- They must be written in good English.
- The topic should fall within the confines of Chemistry, Physics, Earth Sciences, Engineering, and related interdisciplinary fields such as Materials, Nanoscience, Chemical Engineering, Complex Systems, and Biophysics.
- The work reported in the thesis must represent a significant scientific advance.
- If the thesis includes previously published material, permission to reproduce this must be gained from the respective copyright holder.
- They must have been examined and passed during the 12 months prior to nomination.
- Each thesis should include a foreword by the supervisor outlining the significance of its content.
- The theses should have a clearly defined structure including an introduction accessible to scientists not expert in that particular field.

Jean-Daniel Bancal

On the Device-Independent Approach to Quantum Physics

Advances in Quantum Nonlocality and Multipartite Entanglement Detection

Doctoral Thesis Submitted by
the University of Geneva, Switzerland

Author
Dr. Jean-Daniel Bancal
Centre for Quantum Technologies
National University of Singapore
Singapore

Supervisor
Prof. Nicolas Gisin
Group of Applied Physics
University of Geneva
Geneva
Switzerland

ISSN 2190-5053 ISSN 2190-5061 (electronic)
ISBN 978-3-319-37753-7 ISBN 978-3-319-01183-7 (eBook)
DOI 10.1007/978-3-319-01183-7
Springer Cham Heidelberg New York Dordrecht London

Printed on acid-free paper

Springer is part of Springer Science+Business Media (www.springer.com)

Supervisor's Foreword

One of the most fascinating intellectual adventures of humankind started in the 1920s with quantum theory, the theory of atoms and photons (particles of light). Quantum theory has been extremely successful, both in the vast scopes it covers, from very low energies up to nuclear power, from inside atoms and molecules up to the entire cosmos, and in terms of applications like the lasers used in today's DVDs and the semi-conductors used in all modern electronics. Yet, despite 90 years of successes, quantum theory is still poorly understood.

In quantum physics, we describe the measurements we perform and the results we obtain as classical, i.e., not quantum. But then, where is the quantum/classical boundary? Most physicists merely ignore this "measurement" problem. Furthermore, we describe physical systems separated by arbitrary distances as independent, yet quantum theory predicts that they can be correlated even more strongly than is classically possible, allowing for the violation of "Bell inequalities." But, how does Nature do it? Most physicists merely ignore this form of "nonlocality."

Admittedly, some do not ignore these questions, but instead, quite the opposite, they enter into endless and animated debates, as heated as they are vague.

Only recently a community of physicists and computer scientists realized that the classical inputs (measurement settings) and outputs (measurement results) on distant systems with correlations violating some Bell's inequality open entirely new ways to do physics, i.e., to analyze the power of quantum correlations. The only necessary assumption is that distant systems cannot communicate without exchanging some physical messengers carrying the information, quite an obvious assumption, named the no-signaling principle.

Device-Independent Quantum Information Processing (DIQIP), as this new approach to physics is called, is the subject of this Ph.D. thesis. This name reminds one that, in addition to the no-signaling principle, only the input–output relations between classical variables are needed. In particular, neither description nor understanding of the internal functioning of the measurement devices is needed. Think about it. It is truly astonishing that anything nontrivial can be deduced from such minimal assumptions.

This thesis first introduces the mathematical tools needed to study DIQIP. Next, Jean-Daniel Bancal illustrates the power of these tools by analyzing the 2-partite case and—mostly—some multi-partite scenarios. I consider the possibility to detect genuine multi-partite entanglement in this framework as especially remarkable. All these theoretical analyses are accompanied by presentations of experimental results. Finally, the last chapter deals with the profound question "How does Nature produce nonlocal and yet no-signaling correlations?", a question to which Jean-Daniel contributes here with a negative but amazing result … for the reader to discover.

Geneva, April 2013 Prof. Nicolas Gisin

Abstract

During the last century, quantum physics participated in the development of numerous fields: be it computer science which relies on transistors to manipulate information electronically, communications, made possible on large scales thanks to laser light guided by fiber optics, or medicine with the recent development of noninvasive imaging techniques. Who could have forseen that the quantum hypothesis formulated by Max Planck at the dawn of the twentieth century would have such a repercussion?

Nevertheless, quantum physics still remains fairly mysterious. One of its most intriguing aspects being probably its nonlocal character, i.e., the possibility it offers to violate a Bell inequality with systems isolated from each other. Such a violation indeed suggests the existence of a causal connection between admittedly separated systems.

The way in which nonlocality appears in experimental results makes it testable under a minimum set of hypotheses. In particular, no calibration error of individual measurement devices can question the result of such an experiment. This robustness toward implementation errors which are inherent to every experimental realization, opens the way for new experimental approaches. It shows that whenever measured systems are sufficiently separated from each other some questions can be answered by calling upon virtually no additional hypothesis.

Which question can be answered in this way? What can a Bell inequality violation be used for? But also, how does nature manages to violate a Bell inequality? What are the limits of quantum nonlocality? Here are some of the questions considered in this thesis.

Acknowledgments

First of all, I would like to thank Prof. Nicolas Gisin without whom none of the work presented here would have been possible. I am very grateful for the opportunity he gave me to join his group, as well as for his availability for discussions, and in general for his constant support.

A long time ago, Cyril Branciard, Nicolas Brunner, and Christoph Simon accompanied my first steps in the field of quantum information. Thank you!

I owe Stefano Pironio a great deal for all he gave me, from an ounce of mathematical rigor to advices on belgium chocolate.

Thanks to Yeong-Cherng Liang who has always been of great support to me, and whose complementary point of view on many subjects I very much appreciated.

My thanks also go to Nicolas Sangouard for all these 'short' discussions…

It was both a privilege and a pleasure to work with the Innsbruck team. Thanks a lot to Prof. Rainer Blatt for making this possible, and to Julio Barreiro for the correspondences.

I'm very grateful to Enrico Pomarico for sharing with me concerns that a physicist can face during an experiment, and for his company during conferences which I enjoyed a lot.

Thank you Bruno Sanguinetti for this marvelous time in Prague.

I am also thankful to Antonio Acín who invited me several times to Castelldefels, and who suggested subjects to work on with some of his coworkers. Thanks also to Mafalda Almeida and Lars Würflinger for the nice collaborations.

I would also like to thank Tamás Vértesi for the work we did together, as well as the many other visitors who came to the GAP for a day or more of exchange.

Thank you Clara for your cheerfulness.

Thanks also to Michael Afzelius, Denis Rosset, Charles Lim Ci Wen, Tomy Barnea, Basile Grandjean, Pavel Sekatski, Raphael Ferretti-Schöbitz, Markus Jakobi, Keimpe Nevenzeel and all members of the GAP, which I had the chance to meet; thanks for your friendliness.

Finally, I am very grateful to my friends and family for their support. Thank you!

Contents

Chapter 1
Introduction

From its beginning in the 1920s quantum physics has challenged our understanding of the world. Particles that could be conceived previously as points turned out to be provided with a wave evolving in time according to a law of motion. This conceptual change made the observation of previously unsuspected phenomenons possible, like for instance the interference of a molecule with itself demonstrated several times experimentally (e.g. with C_{60} molecules in [1]).

If quantum theory is recognized for its extraordinary predictive power, the picture of the world that it suggests is not the subject of a common agreement. For instance, the question of whether the wavefunction $|\psi\rangle$, a fundamental ingredient of the theory, should be understood as a proper physical object, i.e. a physical property of every quantum system, or rather as a tool from the theory which is only useful to predict the evolution of physically relevant objects, is still an active subject of research [2–5].

One could argue that questions about the possible interpretation of the elements of the quantum theory are of secondary importance, provided that predictions match experimental results. But that would be putting aside the possibility for such considerations to reveal fundamental properties of nature. For instance, the quantum measurement process is commonly understood as an instantaneous change of the wavefunction throughout all space. If this process is indeed instantaneous, and if the wavefunction is a physical object, then measurement of a quantum system is a strongly nonlocal phenomenon, and one should expect physical quantities to be the subject of such instantaneous change at a distance. On the other hand, if the wavefunction can be understood as a tool of the theory, without a concrete physical counterpart, then the nonlocal character of this process might just be an artifact of the theory, without direct incidence on physically relevant quantities.

Since a proper understanding of the elements of the quantum theory seems difficult to reach without invoking arguable choices of additional assumptions, and since only properties of nature that can have a measurable impact are worth discussing anyway, we take the position here to ask what properties of quantum physics can be detected directly from experimental data, without relying on more assumptions than the ones needed in order to make sense of these data. In this way, we hope to be able to explore

J.-D. Bancal, *On the Device-Independent Approach to Quantum Physics*, Springer Theses,
DOI: 10.1007/978-3-319-01183-7_1,

properties of quantum physics more straightforwardly. Moreover, we can expect to be able to check these properties on nature directly, because we follow an approach which fundamentally relies on experimental results.

Following the seminal work of John Bell [6], we consider experimental setups characterized by a number n of identifiable systems, which can be measured in m possible ways, yielding each time one out of k possible values. The results of such a *Bell-type experiment* can be characterized by conditional probability distributions of the form $P(ab|xy)$ (here for a scenario with $n = 2$ parties), which we refer for short as *correlations*. These correlations describe how often the results a and b are observed on two separated systems whenever measurement x and y are performed on them, respectively.

An important property of correlations is that they are always accessible in principle: by sufficiently separating the systems under study, and performing enough measurements, the raw data produced during an experiment allows one to evaluate $P(ab|xy)$ directly. Namely, if the measurements x and y are performed by the two parties Alice and Bob $N(xy)$ times, leading to $N(ab|xy) \leq N(xy)$ observations of the outcomes a and b, then

$$P(ab|xy) = \lim_{N(xy)\to\infty} \frac{N(ab|xy)}{N(xy)}. \tag{1.0.1}$$

Statistical analysis can be used to infer the value of $P(ab|xy)$ with high probability when the number of measurements performed is finite ($N(xy) < \infty$).

Moreover the evaluation of the correlations $P(ab|xy)$ requires no knowledge about the process creating the experimental results. It thus does not rely on any interpretation of the elements of a theory susceptible of describing these processes. Rather, all which is needed in order to make sense out of correlations is well-defined systems and indices to identify the inputs and outputs of the experiment in a reliable fashion. Since this requires no precise description of the working of the measurement devices we refer to it as *device-independent*. This makes correlations well adapted for our purpose. They are thus the central object of interest in this thesis.

Note that apart from allowing one to study nature with a minimum number of assumption, the device-independent hypotheses are also naturally adapted to the study of problems involving untrusted devices, such as quantum key distribution (QKD) [7], or to derive conclusions that are particularly robust with respect to practical uncertainties. While we present a possible application of the second kind in Chaps. 6 and 7, a significant part of this thesis is devoted to the study of correlations in multipartite scenarios.

Before taking a closer look at our contributions, let us recall two fundamental notions that lie behind all of them: the principle of no-signalling and Bell's notion of local causality.

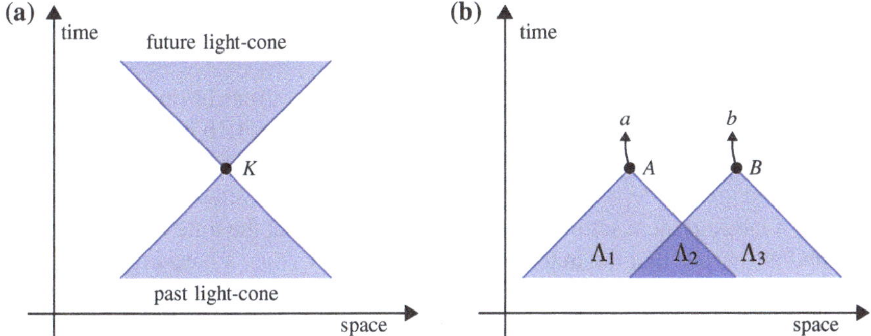

Fig. 1.1 **a** Space-time diagram showing the regions containing events that can influence, or that can be influenced by K according to Bell's principle of local causality. **b** Bipartite Bell experiment in which events creating the outcomes a and b are space-like separated. Apart from the inputs x and y which are chosen at A and B, the outcome a can depend only on the regions Λ_1 and Λ_2, and b only on Λ_2 and Λ_3

1.1 No-signalling and Local Causality

From our everyday experience, we know that any transmission of information (i.e. communication) must be carried by a physical support: in order to let someone know about something we can say it to him, write him an SMS or a letter about it, etc, i.e. use either acoustic waves, electrons, or paper to carry this information to our friend. This idea can be expressed in the following principle:

No-signalling principle. Any transmission of information must be carried by a physical support leaving the emitter after the message is chosen.

This principle is satisfied by several if not all physical theories, including classical and quantum physics [8, 9]. In fact, the no-signalling principle is tightly related to quantum physics, since it can be seen to restrict both the possibility of cloning quantum systems and the possibility of discriminating between quantum states, two peculiarities of quantum physics [10, 11].

In practice, many physical supports are available for communication in nature, like the ones mentioned above, but since the advent of special relativity, it is generally admitted that none of them can carry information faster than light in vacuum. The existence of an upperbound on the speed of all communication led Bell to enunciate the principle of *local causality*:

"The direct causes (and effects) of events are near by, and even the indirect causes (and effects) are no further away than permitted by the velocity of light." J. S. Bell [12]

In other words, the speed of light $c = 299'792'458\,\left[\frac{m}{s}\right]$, taken as an upper bound on any communication speed, naturally defines the limit between events in space-time which can have a direct causal relation with each other, and events which cannot. More precisely, for every event K the set of events that can be influenced by a decision taken at K is defined by its future light-cone. Similarly, all events that can influence a process taking place at K are contained in its past light-cone (c.f. Fig. 1.1a).

1.1.1 Local Correlations

It is of particular interest to ask which correlations $P(ab|xy)$ can be observed, according to the principle of local causality, in the situation of Fig. 1.1b in which two measurements are performed in a space-like manner. Indeed, in this case, no information can be exchanged directly between the two measurement events. Letting the (free) choices of measurement setting x and y be made locally at the time of measurement in A and B, we see that the outcome a is produced by Alice's measurement device before any information about Bob's choice of measurement y can reach it. Thus, a cannot depend on y. Similarly the production of b by Bob's measurement device cannot depend on x.

Still, the measurement processes at A and B can depend on more variables than only x and y. In particular, a is allowed to depend on the whole content of its past light-cone, including Λ_2, a region of space-time which can also influence the process creating b, and Λ_1 which cannot influence b. Let us thus denote by $\lambda_1, \lambda_2, \lambda_3$ all variables which belong to the corresponding regions $\Lambda_1, \Lambda_2, \Lambda_3$ and which are relevant to make predictions about a and b. The most precise prediction of a that can be given prior to measurement and in agreement with the principle of local causality can then be described by a probability distribution of the form $P_A(a|x, \lambda_1, \lambda_2)$. Allowing this distribution to also depend on λ_3, resulting in a prediction of the form $P_A(a|x, \lambda_1, \lambda_2, \lambda_3)$, can only give a more precise description of a than a locally causal theory.[1] Similarly, the distribution $P_B(b|y, \lambda_1, \lambda_2, \lambda_3)$ describes predictions about b at least as well as any locally causal theory.

Since the processes happening at A and B cannot influence each other, they are independent. The average bipartite correlations produced in this situation must thus be of the form

$$P(ab|xy) = \sum_{\lambda} p(\lambda) P_A(a|x, \lambda) P_B(b|y, \lambda) \tag{1.1.1}$$

where $\lambda = (\lambda_1, \lambda_2, \lambda_3)$ and $p(\lambda)$ is a probability distribution, i.e. $p(\lambda) \geq 0$, $\sum_\lambda p(\lambda) = 1$. We refer to this decomposition as the *locality condition*. All correlations which can be decomposed in this way are called *local*, and conversely all correlations which admit no such decomposition are referred to as being *nonlocal*.

Note that here the regions $\Lambda_1, \Lambda_2, \Lambda_3$ in Fig. 1.1b extend up to immemorial times and depend on the precise space-time positions of A and B. Bell showed that different regions Λ' with nicer properties can be chosen in order to reach the same decomposition (1.1.1). Namely, any region Λ' that screens off the regions Λ_i that we considered here, i.e. that already contains the information from Λ which is relevant to make predictions about a and b [13], is good enough to reach Eq. (1.1.1).

[1] Remember that x and y are only chosen at A and B, in a way that is independent of λ_1, λ_2 and λ_3. Allowing a prediction of a to depend on λ_3 thus still doesn't allow it to depend on y.

1.1.2 No-signalling Correlations

If no decomposition of the form (1.1.1) exists for some correlations $P(ab|xy)$, some kind of influence must have taken place between the two measurement events. Yet, this influence might not be available through the correlations to transmit a message. Indeed, users having only access to the variables a, b, x, y can only encode a message to be carried from A to B by the influences in the choice of their inputs x and y. And this message can only be decoded from the observation of the outcomes a and b. Thus, in order to be able to use some correlations to communicate, the statistics of one party's outcome must depend on the other party's choice of measurement. In other words they must violate one of the *no-signalling conditions*:

$$\begin{aligned} P(a|xy) &= \sum_b P(ab|xy) = P(a|x) \ \forall\, y \\ P(b|xy) &= \sum_a P(ab|xy) = P(b|y) \ \forall\, x. \end{aligned} \tag{1.1.2}$$

Note that these constraints are also sufficient: if $P(b|xy) \neq P(b|x'y)$ for some x, x', y, b, then Alice can always send a message to Bob by choosing between x and x', and repeating the experiment enough times to allow for Bob to discriminate between these two probabilities. Correlations satisfying the conditions (1.1.2) are called *no-signalling*.

Violation of the no-signalling conditions allows for communication, which is very common in nature. However, violation of these constraints between space-like separated events would allow for *faster-than-light communication*. Assuming that no physical support can carry information faster than light, this would directly contradict the principle of no-signalling. We come back to this point in the last part of this thesis.

1.1.3 Geometrical Representation

When talking about correlations, it is often useful to represent these probabilities in the vector space obtained by concatenating all components of $P(ab|xy)$. Let us briefly describe a few sets of correlations in this space.

For concreteness, we consider here the scenario where $a, b, x, y = 0, 1$ can only take binary values. Every conditional probability distribution $P(ab|xy)$ can then be represented as the vector

$$\vec{p} = (P(00|00), P(10|00), P(00|10), P(10|10), \ldots, P(11|11)) \in \mathbb{R}^{16} \tag{1.1.3}$$

which belongs to a 16-dimensional vector space. Since probabilities satisfy the normalization condition $\sum_{ab} P(ab|xy) = 1 \,\forall\, x, y$, the space spanned by the correlation vectors $\vec{p}$ is in fact only 12-dimensional. Moreover, probabilities are always positive and must thus satisfy the constraints $P(ab|xy) \geq 0 \ \forall\, a, b, x, y$. This restricts the set of vectors $\vec{p}$ that correspond to valid correlations $P(ab|xy)$ within this 12-dimensional space. Since the number of positivity constraints is finite, the set of valid correlation vectors is a polytope (see Appendix A), which we refer to as the *positivity polytope*.

Similarly, no-signalling correlations are normalized and positive. Moreover they satisfy the no-signalling conditions (1.1.2). These linear conditions define the *no-signalling subspace*, which is of dimension 8 here. The set of all no-signalling correlations is thus the slice of the positivity polytope with this subspace. This is again a polytope (see Appendix A), which is usually called the *no-signalling polytope*.

The set of local correlations, as defined by (1.1.1), can also be described by a polytope in the space of correlations. Indeed it is known [14] that any local correlation $P(ab|xy)$ can be decomposed as a convex combination

$$P(ab|xy) = \sum_{\mu} p(\mu) P_{\mu}(ab|xy), \;\; p(\mu) \geq 0, \; \sum_{\mu} p(\mu) = 1 \tag{1.1.4}$$

of deterministic local strategies $P_{\mu}(ab|xy) = P_A(a|x,\mu)P_B(b|y,\mu) \in \{0, 1\}$. On the other hand any convex combination of deterministic local strategies is also local. The set of local correlations thus corresponds to the convex hull of the deterministic local strategies. Since the number of such strategies is finite, this set is also a polytope, the *local polytope*. Whereas any inequality satisfied by the local polytope is a valid *Bell inequality*, the facets of this polytope are *tight Bell inequalities*.

Finally, it is also useful to characterize the set of *quantum correlations*. These correlations are all the ones which can be obtained by measuring a quantum state ρ with some local measurement operator $M_{a|x}$ and $M_{b|y}$. They can thus always be written as

$$P(ab|xy) = \mathrm{tr}(M_{a|x} \otimes M_{b|y}\rho). \tag{1.1.5}$$

where $\rho \geq 0$, $\mathrm{tr}\rho = 1$, $M_{a|x} \geq 0 \sum_a M_{a|x} = \mathbb{1}$, $M_{b|y} \geq 0 \sum_b M_{b|y} = \mathbb{1}$. The set of quantum correlations is convex, but admits an infinite number of extremal points. It is thus not a polytope. Nevertheless, it can be efficiently characterized by a hierarchy of semi-definite programs [15, 16]. While quantum correlations can violate Bell inequalities, these correlations always satisfy the no-signalling condition (1.1.2). The boundary of this set thus lies between the two preceding sets as represented in Fig. 1.2.

Since quantum correlations can be nonlocal, they can require an exchange of influences between the measurement events. However these influences remains out of reach from us because quantum correlations satisfy the no-signalling condition.

1.1.4 Experimental Loopholes

Knowing that quantum physics can violate the locality condition (1.1.1) is one thing. Verifying that nature violates it is another, which requires the observation of a faithful

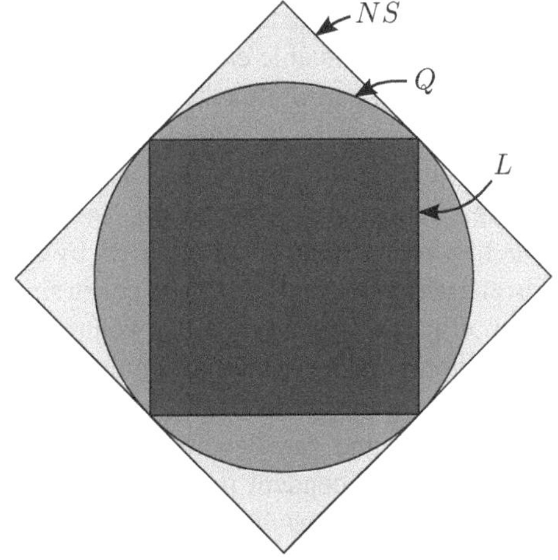

Fig. 1.2 Schematic representation of the set of local (L), quantum (Q), and no-signalling (*NS*) correlations. Note the inclusion $L \subset Q \subset NS$

Bell inequality violation. In particular, such an experiment should demonstrate that no locally causal theory is able to reproduce the experimental results. Given the current technological limitations, all Bell inequality violations demonstrated so far suffer from at least one of the following two loopholes, which prevents them from strictly concluding about the nonlocal character of nature.

The locality loophole. As discussed previously, space-time separation between the measurement events should be guaranteed in order to prevent any communication between the measurement devices. More precisely, one should make sure that the speed of light prevents Alice's choice of measurement setting to reach Bob's device before it produces its outcome. And similarly for Bob's setting. This puts stringent constraints on the timing at which the measurements should be performed, on their possible duration, and on the distance that should separate them.

The detection loophole. If the measurement devices fail to produce outcomes a or b too often, because the systems to be measured are frequently lost along the way for instance, then there is a possibility that discarding the non-detected events allow for a local model to reproduce the post-selected correlations [17]. The probability that the measurements produce results, given some inputs x and y, should thus not be too low.

1.2 Outline

The content of this thesis is organized as follows.

First, we discuss several studies on nonlocality in bipartite scenarios, including a proposal for a loophole-free Bell experiment combining measurements on an atom and a photon, and the analysis of Bell tests in presence of multipairs. We then also present an experimental demonstration of nonlocal correlations conducted with a commercially-available entanglement source.

The second part of this thesis, starting with Chap. 4, discusses the notion of nonlocality in scenarios involving three or more parties. We discuss the definition of genuine multipartite nonlocality, and present a family of inequalities that can detect multipartite nonlocal correlations. In this part we also study more specifically the structure of multipartite correlations by analysing the set of tripartite no-signalling correlations and questioning the constraints that relate different marginals of a single multipartite system. Finally, we provide a bound on the nonlocality of quantum correlations, and a model that simulates measurements on a GHZ state with the help of bipartite nonlocal boxes.

Chapters 6 and 7 are devoted to the detection of genuine multipartite entanglement in a device-independent manner. We examine in which case genuine multipartite entanglement can be witnessed based solely on the observation of some correlations. This allows one to witness multipartite entanglement, a quantum property, in a way that is particularly resistant to practical imperfections. These results are illustrated experimentally. (The theoretical article published on this subject is then reproduced.)

Chapter 8 reminds us that quantum physics allows one to perform some tasks which would be impossible or harder otherwise. It contains the analysis of a specific attack on the 6-state QKD protocol, as well as a proposal for practical secure database queries.

Finally, we close this thesis by considering in Chaps. 9 and 10 the possibility of relaxing Bell's condition of local causality to recover a causal explanation of quantum nonlocal correlations with a sense of proportion. We show that this is not possible without allowing for faster-than-light communication.

The text presented here alternates between original chapters and selected articles, reproduced as Chaps. 3, 5, 7 and 10. While the text in the original chapters is meant to be concise, complementary information can be found in the reproduced papers or online at http://www.arXiv.org. The different parts of this thesis can be read independently of each other.

References

1. M. Arndt, O. Nairz, J. Vos-Andreae, C. Keller, G. van der Zouw, A. Zeilinger, Nature **401**, 680–682 (1999)
2. M.F. Pusey, J. Barrett, T. Rudolph, Nat. Phys. **8**, 476 (2012)
3. M.J.W. Hall, TGeneralisations of the recent pusey-barrett-rudolph theorem for statistical models of quantum phenomena. Quantum Phys. (Submitted on 27 Nov 2011) arXiv:1111.6304
4. R. Colbeck, R. Renner, Phys. Rev. Lett. **108**, 150402 (2012)
5. H.F. Hofman, The quantum state should be interpreted statistically. Quantum Phys. (Submitted on 12 Dec 2011) arXiv:1112.2446
6. J.S. Bell, *Speakable and Unspeakable in Quantum Mechanics* (Cambridge University Press, Cambridge, 1987)
7. Ll. Masanes, S. Pironio A. Acín, Nat. Comm. **2**, 238 (2011)

8. M. Pawłowski, T. Paterek, D. Kaszlikowski, V. Scarani, A. Winter, M. Żukowski, Nature **461**, 1101–1104 (2009)
9. T. Mauldin, *Quantum Non-Locality and Relativity: Metaphysical Intimations of Modern Physics* (Blackwell, Oxford, 2002)
10. N. Gisin, Phys. Lett. A **242** (1998)
11. J. Bae, W.-Y. Hwang, Y.-D. Han, Phys. Rev. Lett. **107**, 170403 (2011)
12. J.S. Bell, La nouvelle cuisine, *Speakable and Unspeakable in Quantum Mechanics*, 2nd edn. (Cambridge University Press, Cambridge, 2004)
13. T. Norsen, J.S. Bell's Concept of Local Causality. Am. J. Phys. **79**, 1261 (2011) arXiv:0707.0401
14. I. Pitowsky, Brit. J. Phil. Sci. **45**, 95 (1994)
15. M. Navascués, S. Pironio, A. Acín, New J. Phys. **10**, 073013 (2008)
16. Y.-C. Liang, A.C. Doherty, Phys. Rev. A **75**, 042103 (2007)
17. N. Gisin, B. Gisin, Phys. Lett. A **260**, 323 (1999)

Chapter 2
Bell Tests in Bipartite Scenarios

2.1 Bell Test Between an Atom and an Optical Mode

If Bell experiments conducted so far have always suffered from one of the loophole described above, technological advances suggest that both the locality and the detection loophole might soon be closable within the same experiment. In order to make this happen, novel proposals taking into account present capabilities are highly welcome. Here we describe a proposal for a loophole-free Bell test, and analyse its feasibility.

Bell tests with photonic systems are well designed to ensure strict space-like separation between the measurement events, thanks to the high speed at which photons can travel. However optical losses are unavoidable, leaving the detection loophole open. On the other hand, atomic systems can provide very high detection efficiency, but don't travel well enough to allow for a space-time separation between the measurements. To close both loopholes, we consider here an hybrid entangled system consisting of an atom (which can be detected very efficiently) and a photon (which can travel fast, and thus helps to close the locality loophole).

We first describe how entanglement between an atom and a photon can be produced, and then discuss the constraints that an experiment would have to satisfy in order to allow an experiment on this system to demonstrate nonlocality.

2.1.1 Creating Atom-Photon Entanglement

Let us consider an atom with a lambda-type level configuration (as depicted in Fig. 2.1), initially prepared in the state $|g\rangle$. A pump laser pulse with the Rabi frequency Ω can be used to partially excite the atom in such a way that it can spontaneously decay into the level $|s\rangle$ by emitting a photon. Long after the decay time of the atom, the atom-photon state is given by

DOI: 10.1007/978-3-319-01183-7_2,

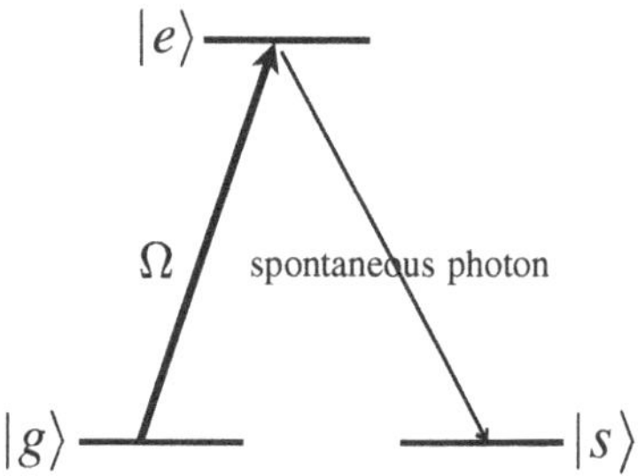

Fig. 2.1 Basic level scheme for the creation of atom-photon entanglement by partial excitation of an atom. The branching ratio is such that when the atom is excited, it decays preferentially in $|s\rangle$

$$|\psi^{\phi}\rangle = \cos(\theta/2)|g, 0\rangle + e^{i\phi}\sin(\theta/2)|s, 1\rangle \tag{2.1.1}$$

where $\theta = \int dt\Omega(t)$ refers to the area of the pump pulse. The phase term is defined by $\phi = k_p r_p - k_s r_s$ where k_p (k_s) corresponds to the wave vector of the pump (the spontaneous photon) and r_p (r_s) is the atom position when the pump photon is absorbed (the spontaneous photon is emitted).

2.1.2 CHSH Violation

In order to demonstrate nonlocality with the above state, we propose to test the CHSH Bell inequality [1]:

$$S = E_{00} + E_{01} + E_{10} - E_{11} \leq 2 \tag{2.1.2}$$

where $E_{xy} = p(a = b|xy) - p(a \neq b|xy)$ is the correlation between Alice and Bob's outcomes when they respectively perform measurements x and y.

Here we consider that Alice can choose measurement bases for her qubit on the whole Bloch sphere. However, since Bob's qubit lies in the Fock space spanned by 0 and 1 photons, we let him only choose between two kinds of natural measurements: photon counting and homodyne measurements.

Since measurements on the atom can be very efficient, we assume that they always produce an outcome. Similarly, homodyne measurements can be very efficient [2] so that Bob's homodyne measurement is considered perfectly efficient. However, we let his photon counter have a detection efficiency η_d: when a photon arrives on his detector, it thus produces a click with probability η_d.

To analyse the impact of the distance between Alice and Bob, we model the channel through which the photon propagates as a lossy channel with transmission η_t:

$$\begin{aligned} &|0, 0\rangle \rightarrow |0, 0\rangle \\ &|1, 0\rangle \rightarrow \sqrt{1-\eta_t}|0, 1\rangle + \sqrt{\eta_t}|1, 0\rangle \end{aligned} \tag{2.1.3}$$

here the second qubit is a mode of the environment, which is not observed. Tracing out this mode, we get an effective state after the transmission line of

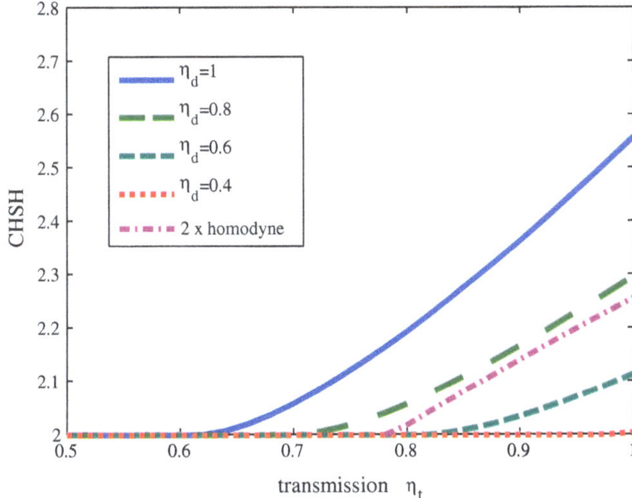

Fig. 2.2 Amount of CHSH violation achievable in an atom-photon Bell experiment. The *dashed-dotted line* corresponds to the case in which both measurements on the photonic mode are homodyne measurement. The other *curves* are for one homodyne measurement and a photon counting. The lowest permissible transmission here is $\eta_t = 61\,\%$ and the lowest photo-detection efficiency is $\eta_d = 39\,\%$

$$\rho_{\eta_t} = (\cos\theta|g,0\rangle + e^{i\phi}\sin\theta\sqrt{\eta_t}|s,1\rangle)(\cos\theta\langle g,0| + e^{-i\phi}\sin\theta\sqrt{\eta_t}\langle s,1|) + (1-\eta_t)\sin^2\theta|s,0\rangle\langle s,0|. \quad (2.1.4)$$

Considering this state and Eq. (2.1.2) together, we optimized the free parameters in the state and measurements to get the largest violation for several choices of η_t and η_d. The result is plotted in Fig. 2.2.

2.1.3 Space-like Separation

From Fig. 2.2, we see that the test above provides a certain robustness with respect to losses and detection inefficiency. In order to close both loopholes, these quantities should be compared with the losses expected from an experiment ensuring space-like separation of the measurements. These are ultimately determined by the time needed in order to perform the measurements on the atom or the photon.

In our case, we expect the slowest measurement to be the atomic one. Still, the measurement should take about 1 μs [3]. A distance of the order of 300 m would thus be needed to ensure space-like separation. For 800 nm photons, a fiber of this length has a transmission of 93 %. The scheme with double homodyne measurements is compatible with these requirements (see Fig. 2.2), which attests of the potential feasibility of this experiment.[1]

2.1.4 Conclusion

We showed that a sensible violation of the CHSH inequality can be obtained by combining measurements on an atom with photon counting and homodyne measurements on an optical mode. We also argued that the discussed quantum state could be produced with existing technology.

Any practical implementation of the above scheme would involve imperfections. For instance, the branching ratio of the atom may not be perfect, meaning that the excited level $|e\rangle$ in Fig. 2.1 could decay to other levels than $|s\rangle$, which we didn't take into account here. The movement of the atom during the application of the Ω pulse can have an influence on the phase ϕ of the produced state as well, and the transmission line of the photon should be stable enough not to loose this phase. All of these aspects can be shown not to threaten directly the main conclusion (see [4] for more details). This supports the idea that a Bell experiment closing both the locality and the detection loophole is close to technological reach (see also [5, 6] for more proposals along these lines).

2.2 Bell Test with Multiple Pairs

Bell experiments are often realized by measuring one pair of particles at a time. However, there are situations in which the entanglement produced is shared by many particles which cannot be addressed individually. For instance, in [7], many pairs of entangled ultracold atoms have been produced, but they cannot be measured individually. One can thus wonder whether the violation of a Bell inequality could in principle be tested in such many-body systems.

Here we consider systems consisting of M particles, which cannot be addressed individually. The measurements thus act identically on all the particles that each part receives. For encoding in polarization, this can be modeled by a polarizing beam-splitter (PBS) followed by photon counters (c.f. Fig. 2.3). In this situation, the detectors following the PBS can receive different numbers of particles. In order to recover binary outcomes allowing to test the CHSH Bell inequality [1], we introduce a post-processing of the outcomes. Namely, whenever the number of photons detected in the '+' port n_+ is greater than or equal to a given threshold N, the outcome is set to '+1', otherwise it is set to '−1'. Two particular strategies of interest here are the majority voting strategy when $N = M/2$, and the unanimity voting strategy when $N = M$.

---

[1] Note however that the total transmission efficiency also includes the collection efficiency, i.e. the probability with which the photon emitted by the atom is collected into a fibre. Collection efficiencies of the order of 50 % have already been demonstrated using a cavity (see [4]).

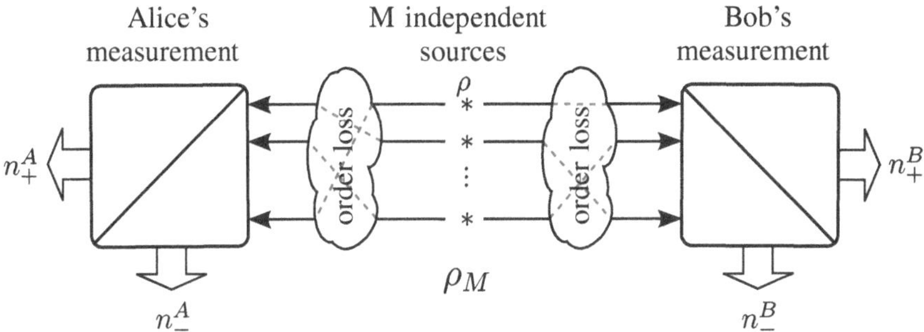

Fig. 2.3 Setup of a multipair Bell experiment: here a source produces M independent pairs of entangled particles. Since the pairing between Alice's and Bob's particles is lost during their transmission, all particles are measured identically by each party. The total number of particles detected in both outcomes + and − are tallied on both sides

2.2.1 Two Sources

Within this measurement setup, we consider two possible sources of entangled particles: a source of distinguishable particles, and a source of indistinguishable ones. The first one produces states of the form

$$\rho_M = \rho^{\otimes M} = (|\psi\rangle\langle\psi|)^{\otimes M} \tag{2.2.1}$$

where $|\psi\rangle = \frac{1}{\sqrt{2}}(|00\rangle + |11\rangle)$. The second source produces indistinguishable particles in the state

$$|\Phi_M\rangle = \frac{1}{M!\sqrt{M+1}}(a_0^\dagger b_0^\dagger + a_1^\dagger b_1^\dagger)^M|0\rangle. \tag{2.2.2}$$

States of this second form can be obtained for instance by parametric down conversion (PDC), which produce Poissonian distributions of such states.

2.2.2 Noise Model

In order to quantify the amount of CHSH violation for each source, we introduce a noisy channel between the source and Alice. This channel consists of a random unitary $U = \exp(-\beta\vec{n}\cdot\vec{\sigma})$ applied to the state, where the rotation axis $\vec{n}$ is uniformly distributed on the Bloch sphere, and the angle β follows a gaussian distribution $p(\beta) = \frac{2}{(1-e^{-2\sigma^2})\sqrt{2\pi}\sigma}e^{-\frac{\beta^2}{2\sigma^2}}$ centered at the origin and of variance σ^2. The state after this channel is given by

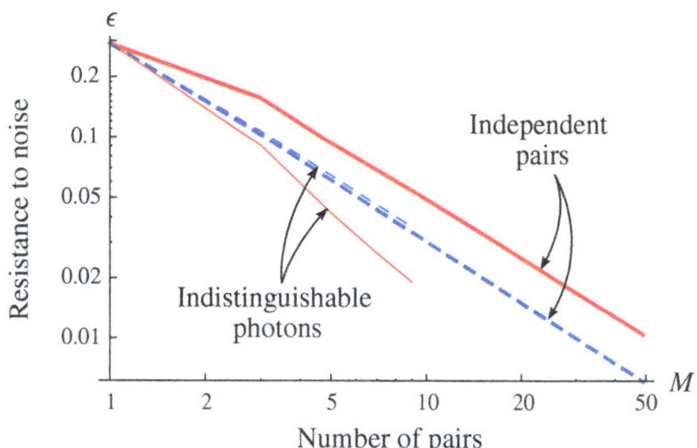

Fig. 2.4 Maximal resistance to noise in the majority voting scenario (*full red lines*) and the unanimity scenario (*dashed blue lines*) for sources producing independent pairs (*thick lines*) or indistinguishable photons (*thin lines*). The unanimous vote is more robust with indistinguishable photons, but majority voting on independently produced pairs yields the most persistent violation

$$\rho_{out} = \int_{SU(2)} p(\beta)(U \otimes \mathbb{1})\rho_{in}(U^{\dagger} \otimes \mathbb{1})[dU] \tag{2.2.3}$$

where $[dU]$ is the Haar measure on SU(2).

The application of this channel on the maximally-entangled state $|\Phi_1\rangle = \frac{1}{\sqrt{2}}(a_0^{\dagger}b_0^{\dagger} + a_1^{\dagger}b_1^{\dagger})|0\rangle$ produces the Werner state [8]

$$\rho = w|\Phi_1\rangle\langle\Phi_1| + (1-w)\frac{\mathbb{1}}{4} \tag{2.2.4}$$

with $w = \frac{1}{3}\left(e^{-2\sigma^2} + e^{-4\sigma^2} + e^{-6\sigma^2}\right)$. We thus quantify violations by the largest amount of noise $\epsilon = 1 - w \simeq 4\sigma^2$ which still allows one to find a violation of the CHSH inequality.

2.2.3 Bell Violation

The best resistance to noise found after optimizing on the settings is represented as a function of the number of photons produced M in Fig. 2.4. Interestingly, both sources provide a finite violation even for fairly large numbers of particles. Still, the maximum violation decreases like M^{-1} in both cases, in connivance with the principle of macroscopic locality [9]. Thus, if a Bell violation can be found in multipair systems, it becomes less and less significant as the number of particles involved increases.

2.3 Experimental Violation of Bell Inequalities with a Commercial Source of Entanglement

Back in the 1970–1980s, the first experimentalists to test Bell inequalities had to put special efforts in building sources of entangled particles [10]. Since then, a lot of effort has been done to improve these sources. Today, it is possible to buy sources of entangled particles that are ready to test a Bell inequality. Here we demonstrate the violation of several Bell inequalities that we obtained with a commercially available source.

2.3.1 Experimental Setup

We used the QuTools source [11] to produce pairs of 810 nm photons entangled in polarization via spontaneous parametric down-conversion (SPDC). The source consists of a bulk β-barium borate (BBO) crystal, cut for Type II phase-matching, which is pumped at 405 nm by a continuous wave diode laser (see Fig. 2.5). The two photons produced by the source have orthogonal polarization and are emmited in cones. After selection of the spatial mode corresponding to the intersection of the two cones with pinholes and single-mode fibers, the photons collected can be described by the state

$$|\psi\rangle = \frac{1}{\sqrt{2}}\left(|H\rangle_s|V\rangle_i + e^{i\phi}|V\rangle_s|H\rangle_i\right). \tag{2.3.1}$$

The setup allows us to measure each photon along any direction lying on the equator of the Bloch sphere. A partial tomography in this x-y plane shows that the state is close to a Werner state

$$\rho = V|\psi^-\rangle\langle\psi^-| + (1-V)\frac{\mathbb{1}}{4} \tag{2.3.2}$$

with visibility $V = 94\,\%$.

2.3.2 Test of Several Bell Inequalities

Using this source, we tested the CHSH, I_{3322}, AS_1 and AS_2 Bell inequalities [1, 12–14]. The measurement settings were optimized in each case according to the state knowledge obtained through the partial tomography.

The results are represented in Table 2.1. The values found are in good agreement with the values expected from the partial tomographic knowledge of the source.

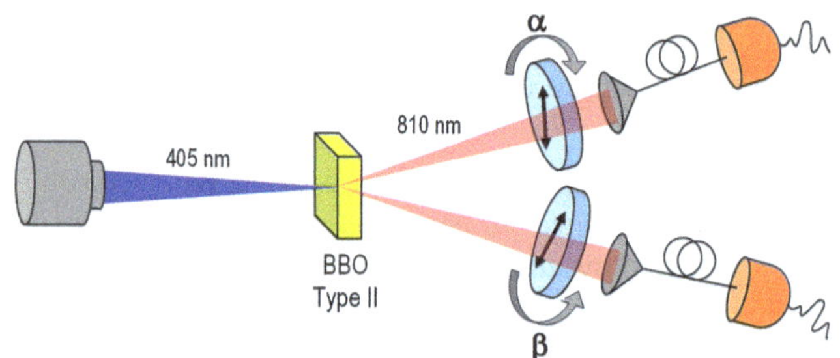

Fig. 2.5 Sketch of the experimental setup used to test various Bell inequalities. Alice's and Bob's choice of settings are adjusted by rotation of linear polarizers

Table 2.1 Measurement of the CHSH inequality and of inequalities inequivalent to CHSH

	I_L	I^{exp}	I^{tom}	$I^{exp} - I_L$ (σ units)	p_{noise} (%)
I_{CHSH}	2	2.731 ± 0.015	2.683	49	27
I_{3322}	4	4.592 ± 0.024	4.769	25	13
AS_1	6	7.747 ± 0.026	7.750	67	23
AS_2	10	12.85 ± 0.030	12.819	95	22

I_L is the local bound, I^{exp} is the value of the Bell parameter obtained experimentally with the optimized settings, I^{tom} is the expected value from the partial tomography, $I^{exp} - I_L$ is the difference between the obtained value and the local bound in terms of number of standard deviations σ and $p_{\text{noise}}(\%)$ is the critical level of white noise that can be added to the system without loosing a violation

Table 2.2 Measurement of the chained inequalities with N settings per side

N	I_L	I^{exp}	I^{tom}	$I^{exp} - I_L$ (σ units)	$p_{\text{noise}}(\%)$
2	2	2.731 ± 0.015	2.683	49	27
3	4	4.907 ± 0.019	4.925	48	18
4	6	7.018 ± 0.023	6.999	44	15
5	8	8.969 ± 0.026	8.996	37	11
6	10	10.91 ± 0.028	10.954	33	8

2.3.3 Chained Bell Inequality

On top of these inequalities, we also tested the N-settings chained Bell inequality, which can be written as

$$I^N = E_{11} + E_{12} + E_{22} + \cdots + E_{NN} - E_{N1} \leq 2(N-1) = I_L^N. \qquad (2.3.3)$$

The values obtained experimentally are reported for $N \leq 6$ in Table 2.2. The chained inequality has a number of applications which we mention below.

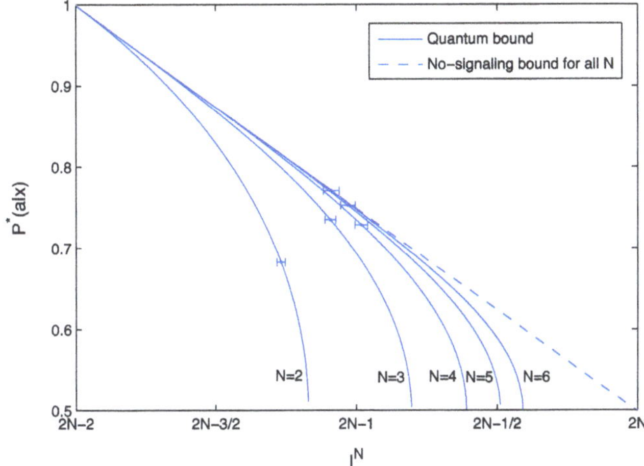

Fig. 2.6 Maximum marginal probability compatible with a violation of the N-settings chained inequality. The bound implied by the no-signalling principle is identical for all N

2.3.3.1 Randomness Certified by the No-signalling Principle

An interesting property of the chained Bell inequality is that the marginal probabilities $P(a|x)$ and $P(b|y)$ tend to $1/2$ as the violation of the inequality increases. This allows one to certify that the outcomes produced by measuring the quantum system must be truly random, in the sense that no algorithm can possibly predict the measurement outcomes [15].

More precisely, the amount of true randomness that could be extracted from the experimental results found by Alice can be evaluated by finding the largest marginal probability $P^*(a|x)$ which is compatible with the measured Bell inequality violation I^{exp}. We performed this optimization over the set of quantum correlations as well as among all no-signalling correlations. The result is shown in Fig. 2.6 together with the experimentally achieved values I^{exp}.

The strongest bound imposed by the no-signalling principle is $P^*(a|x) = 0.7455 \pm 0.0057$, achieved for the inequality with $N = 4$ settings. This allows one in principle to extract $H_{\min}(a|x) = -\log_2 P^*(a|x) = 0.41 \pm 0.01$ random bits per run.

2.3.3.2 EPR2 Local Part

Another property of the chained inequality is that its maximum quantum value $I_Q^N = 2N\cos\left(\frac{\pi}{2N}\right)$, achievable by measuring a singlet state [16], approaches the no-signalling bound $I_{NS}^N = 2N$ as the number of settings N increases. This allows one to conclude that the singlet state has no local part in the sense of EPR2 [17].

Indeed, if a fraction p_L of the measured pairs would behave locally during an experimental evaluation of the chained inequality (2.3.3) yielding the value I^{exp}, the following equation would hold:

$$I^{exp} = p_L I_L^N + (1 - p_L) I_{NL}^N, \tag{2.3.4}$$

where I_L^N is the value of I^N achieved with the local pairs of particles, and I_{NL}^N a value of the same expression achieved on the rest of the particles. Since the following bounds hold: $I_L^N \leq 2(N-1)$ and $I_{NL}^N \leq I_{NS}^N = 2N$, the local part p_L of the measured states must be bounded by

$$p_L \leq p_L^{max} = N - \frac{I^{exp}}{2}. \tag{2.3.5}$$

For $I^{exp} = I_Q^N$, we find $p_L^{max} = N\left(1 - \cos\left(\frac{\pi}{2N}\right)\right) \overset{N\to\infty}{\longrightarrow} 0$. Thus, for every number of settings N, testing the chained inequality can provide an upperbound on the local content of the state measured which eventually converges to 0.

In our case, the best bound on p_L is found for $N = 4$ settings, yielding $p_L^{max} = 0.491 \pm 0.012$. While recent work could demonstrate an even lower value [18], this simple experiment already shows that at least half of the photon pairs produced by the source are nonlocal.

2.3.4 Conclusion

In this experiment, we relied on the fair sampling assumption because the single photon detectors were not efficient enough to close the detection loophole. Moreover the detection events were not space-like separated. Yet, this experiment shows that a simple demonstration of several interesting results of quantum information theory is nowadays possible with modest equipment.

References

1. J.F. Clauser, M.A. Horne, A. Shimony, R.A. Holt, Phys. Rev. Lett. **23**, 880 (1969)
2. H. Nha, H.J. Carmichael, Phys. Rev. Lett. **93**, 020401 (2004)
3. F. Henkel, M. Krug, J. Hofmann, W. Rosenfeld, M. Weber, H. Weinfurter, Phys. Rev. Lett. **105**, 253001 (2010)
4. J.-D. Bancal, N. Gisin, Y.-C. Liang, S. Pironio, Phys. Rev. Lett. **106**, 250404 (2011)
5. A. Cabello, J.-A. Larsson, Phys. Rev. Lett. **98**, 220402 (2007)
6. D. Cavalcanti, N. Brunner, P. Skrzypczyk, A. Salles, V. Scarani, Phys. Rev. A **84**, 022105 (2011)
7. M. Anderlini, P.J. Lee, B.L. Brown, J. Sebby-Strabley, W.D. Phillips, J.V. Porto, Nature **448**, 452–456 (2007)
8. R.F. Werner, Phys. Rev. A **40**, 4277 (1989)

9. M. Navascués, H. Wunderlich, Proc. Roy. Soc. Lond. A **466**, 881–890 (2009)
10. A. Aspect, http://arxiv.org/abs/quant-ph/0402001
11. P. Trojek, C. Schmid, M. Bourennane, H. Weinfurter, C. Kurtsiefer, Opt. Expr. **12**, 276 (2004)
12. D. Collins, N. Gisin, J. Phys. A: Math. and Gen. **37**, 1775 (2004)
13. N. Brunner, N. Gisin, Phys. Lett. A **372**, 3162 (2008)
14. D. Avis, H. Imai, T. Ito, J. Phys. A: Math. and Gen. **39**, 11283 (2006)
15. S. Pironio, A. Acín, S. Massar, A. Boyer de la Giroday, D.N. Matsukevich, P. Maunz, S. Olmschenk, D. Hayes, L. Luo, T.A. Manning, C. Monroe, Nature **464**, 1021–1024 (2010)
16. S. Wehner, Phys. Rev. A **73**, 022110 (2006)
17. A. Elitzur, S. Popescu, D. Rohrlich, Phys. Lett. A **162**, 25 (1992)
18. L. Aolita, R. Gallego, A. Acín, A. Chiuri, G. Vallone, P. Mataloni, A. Cabello, Phys. Rev. A **85**, 032107 (2012)

Chapter 3
Various Quantum Nonlocality Tests with a Commercial Two-photon Entanglement Source

Nonlocality is a fascinating and counterintuitive aspect of nature, revealed by the violation of a Bell inequality. The standard and easiest configuration in which Bell inequalities can be measured has been proposed by Clauser- Horne-Shimony-Holt (CHSH). However, alternative nonlocality tests can also be carried out. In particular, Bell inequalities requiring multiple measurement settings can provide deeper fundamental insights about quantum nonlocality, as well as offering advantages in the presence of noise and detection inefficiency. In this paper we show how these nonlocality tests can be performed using a commercially available source of entangled photon pairs. We report the violation of a series of these nonlocality tests (I_{3322}, I_{4422}, and chained inequalities). With the violation of the chained inequality with 4 settings per side we put an upper limit at 0.49 on the local content of the states prepared by the source (instead of 0.63 attainable with CHSH). We also quantify the amount of true randomness that has been created during our experiment (assuming fair sampling of the detected events).

3.1 Introduction

Nonlocality is one of the most counterintuitive and fascinating aspects of Nature revealed by the quantum theory. Indeed, the fact that two separated systems appear to work in a joint way, independently of the distance separating them, does not have a counterpart in the classical world. In particular, this bizarre effect is predicted by quantum mechanics for two entangled systems and manifests itself in the correlations of the outcomes of the measurements performed on the two systems. This kind of "spooky actions at a distance" was the argument that Einstein, Podolski and Rosen used to claim the incompleteness of the quantum mechanics [1]. They pointed out that one could restore locality in physics by assuming the existence of local variables determining the results of the measurements. However, in 1964 John Bell showed that the correlations of the results of the measurement on entangled particles are stronger

J.-D. Bancal, *On the Device-Independent Approach to Quantum Physics*, Springer Theses, DOI: 10.1007/978-3-319-01183-7_3, © Springer International Publishing Switzerland 2014

with respect to what is expected by any physical theory of local hidden variables. By measuring them, a simple inequality (i.e. the Bell inequality [2]), based on the locality assumption, can be violated.[1]

Testing experimentally nonlocal correlations predicted by quantum mechanics has become a concrete idea after the clauser-horne-shimony-holt (CHSH) formulation [3] of the original Bell inequality. Since the CHSH-Bell tests performed by Aspect in 1981 [4], several experiments have confirmed consistent violations of the CHSH inequality. However, the experimental imperfections, which all these tests are susceptible to, open loopholes that can be exploited by a local theory to reproduce the experimental data. Two relevant loopholes are the detection and the locality loophole. The former relies on the fact that particles are not always detected in both channels of the experiment. The latter is related to the necessity of separating the two sites enough to prevent any light-speed communication between them from the time measurement settings are set until the detection events have occurred. At the moment both loopholes have been closed [5–9], but not yet within the same experiment.

Nowadays, performing nonlocality tests with entangled photons is relatively easy. Indeed, the technology necessary for generating entanglement is well known and entanglement sources can be extremely compact, cheap and stable in time. Moreover, polarization entanglement can be analyzed in a practical and accurate way. Therefore, the experimental study of entanglement is not confined to a few privileged laboratories, but can be also carried out as part of undergraduate laboratory courses.

The CHSH inequality requires a rather simple configuration: two binary measurement settings on each of two entangled particles. However, in recent years new types of Bell inequalities, involving a larger number of measurement settings or outcomes with respect to CHSH, have been investigated from a theoretical point of view. These inequalities can get conclusions unattainable with CHSH: from a fundamental point of view, they allow for a deeper understanding of the nonlocal correlations, whereas from a practical one, they represent useful tools in the presence of noise and detection inefficiency [10], as we show in the next paragraph.

In this paper, we show that, with a commercial source of photon pairs entangled in polarization (QuTools [11]), a series of nonlocality tests alternative to CHSH can be made. In particular, we report the violation of Bell inequalities requiring multiple binary measurement settings. To our knowledge these inequalities (with the exception of I_{3322} [12]) have not been measured before.

In Sect. 3.2 we justify the interest in this kind of inequalities from a fundamental and practical point of view. In Sect. 3.3 we describe the partial tomography of the density matrix of the states prepared by the source and the optimization of the settings for enhancing the violation of the Bell inequalities. In Sect. 3.4 we report the violation of inequalities inequivalent to CHSH, in particular the I_{3322} [13] and two I_{4422} inequalities [14]. Then, we show the violations of chained inequalities [15] from 3 to 6 settings per side. In Sect. 3.5.1 we interpret the violation of the chained inequalities

[1] This chapter appeared as: "E. Pomarico, et al., *Various quantum nonlocality tests with a commercial two-photon entanglement source*, Phys. Rev. A **83**, 052104 (2011)."

according to a specific nonlocality approach introduced by Elitzur, Popescu, and Rohrlich (EPR2) [16]. This allows us to put an upper limit on the local content of the prepared states that is stronger than the one attainable by CHSH. Finally, we show that the observed violations allow one to certify that true random numbers have been created during the experiment. Throughout this work we assume fair sampling of the detected events, which allows us to avoid detection loophole issues.

3.2 Bell Inequalities with Multiple Measurement Settings

A CHSH-Bell test requires the measurement of each photon of an entangled pair in two different bases and the estimation of the correlations in the four possible combinations of bases. Its practical implementation is conceptually easy and needs minimum experimental effort with respect to other nonlocality tests. Moreover, the CHSH test is particularly robust against the noise present in real experiments. However, other Bell inequalities, requiring a larger number of measurement settings, schematically represented in Fig. 3.1, can lead to conclusions about the quantum correlations that are non trivial and sometimes are inaccessible to CHSH. In general, finding all the Bell inequalities for a given setup is computationally hard, therefore the research is limited to a small number of settings per side. Even in this case, these tests can provide interesting insights about quantum nonlocality. In the case of three possible 2-outcome measurements per side, only one inequality, called I_{3322}, is inequivalent to CHSH [13]. Note that this inequality is relevant in the sense that it can be violated by specific mixed 2-qubit states that do not violate CHSH. The I_{3322} inequality has also been used to show that three qubits can share bipartite non-locality between more than two subsystems, a result that cannot be obtained with CHSH [13]. In the case of four 2-outcome measurements, only a partial list of inequalities I_{4422} has been given [14]. Some of these inequalities are maximally violated, surprisingly, by non-maximally entangled states, unlike CHSH.

Another set of inequalities that have recently attracted attention is represented by the chained inequalities [15], which are generalizations of CHSH with multiple settings. In recent years, several theoretical models have attempted to provide a better understanding of quantum nonlocality. One of these is the Elitzur, Popescu, and Rohrlich (EPR2) approach [16], according to which the observed data could

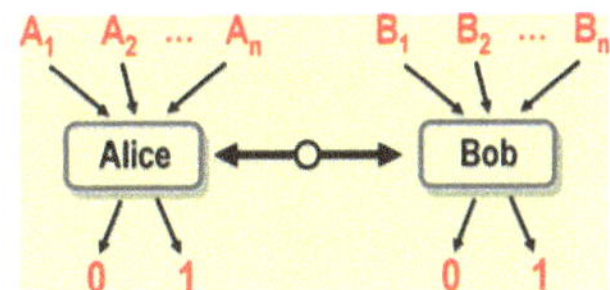

Fig. 3.1 Nonlocality tests where Alice and Bob measure the N binary operators $\{A_1, \ldots, A_N\}$ and $\{B_1, \ldots, B_N\}$ respectively on the photon pairs produced by the entanglement source in the middle

be explained assuming that only a fraction of the photon pairs produced in an experiment possess nonlocal properties, while the remaining part gives rise to purely local correlations. In this scenario chained inequalities are a useful tool for studying the local content of quantum correlations. For instance, they have been used to prove that maximally entangled quantum states in arbitrary dimensions have a zero local component [17] and to decrease the upper bound on the local content of nonmaximally entangled states [18].

From a more practical point of view, inequalities based on multiple settings are also interesting. Recently, it has been shown that in the presence of high dimensional entanglement, that is when the quantum systems sharing entanglement have dimensions larger than two, these inequalities can tolerate a detection efficiency of 61.8 % for closing the detection loophole [10]. This value is lower with respect to the limit imposed by CHSH in experiments of entangled qubits.

3.3 Optimization of the Measurement Settings for a Specific State

Entangled states prepared in the laboratory (or by a commercial source) are not perfect in terms of purity and degree of entanglement. In this case the measurement settings needed to observe the largest possible violation of a given inequality do not necessarily coincide with those that are optimal for a maximally entangled state. In order to find these best measurement settings, a knowledge of the state is required. Usually, a complete reconstruction of the state is not possible or even not necessary. The QuTools source set-up projects the photons only onto linear polarization states (c.f. description of the experimental setup in the next section), so we can only perform a partial tomography of the state. Once we know the state on the equatorial plane of the Poincaré sphere corresponding to linear polarizations, we can optimize the measurement settings in this plane.

3.3.1 Partial Tomography of a Quantum State

The density matrix ρ of a two-qubit state can always be written in the basis composed by the identity $\mathbb{1}$ and the Pauli matrices $\{\sigma_x, \sigma_y, \sigma_z\}$ as

$$\rho = \frac{1}{4}\Big(\mathbb{1} \otimes \mathbb{1} + \sum_{i=x,y,z} a_i \sigma_i \otimes \mathbb{1} + \sum_{j=x,y,z} b_j \mathbb{1} \otimes \sigma_j + \sum_{i,j=x,y,z} c_{i,j} \sigma_i \otimes \sigma_j\Big), \tag{3.3.1}$$

where $a_i = \langle \sigma_i \otimes \mathbb{1} \rangle_\rho$, $b_j = \langle \mathbb{1} \otimes \sigma_j \rangle_\rho$ and $c_{i,j} = \langle \sigma_i \otimes \sigma_j \rangle_\rho$ are 15 real coefficients which completely define the state. A complete tomography [19] allows to determine the value of all these coefficients.

In our case we only measure linear polarizations, so a complete knowledge of the state is not necessary in order to predict all possible measurement statistics. Indeed, only the 8 coefficients $a_i, b_j, c_{i,j}$ with $i, j = x, z$ are useful. These coefficients can be measured with our setup, realizing a partial tomography of the generated state.

3.3.2 Optimization of the Settings

The measurement of a qubit along a particular angle θ in the xz plane of the Bloch sphere can be represented by the measurement operator

$$O(\theta) = \cos\theta\sigma_z + \sin\theta\sigma_x. \tag{3.3.2}$$

If the two-qubit state 3.3.1 is shared between Alice and Bob, the following marginal values are expected if Alice measures along angle α and Bob along β:

$$E(\alpha) = \langle A(\alpha) \otimes \mathbb{1} \rangle_\rho = \cos\alpha a_z + \sin\alpha a_x, \tag{3.3.3}$$

$$E(\beta) = \langle \mathbb{1} \otimes B(\beta) \rangle_\rho = \cos\beta b_z + \sin\beta b_x. \tag{3.3.4}$$

Moreover, the joint correlations are found to be

$$\begin{aligned} E(\alpha, \beta) &= \langle A(\alpha) \otimes B(\beta) \rangle_\rho \\ &= \cos\alpha\cos\beta c_{z,z} + \cos\alpha\sin\beta c_{z,x} \\ &\quad + \sin\alpha\cos\beta c_{x,z} + \sin\alpha\sin\beta c_{x,x}. \end{aligned} \tag{3.3.5}$$

A Bell inequality I with N settings per side [20] can be generally defined by the following formula

$$I = \sum_{i=1}^{N} n_i E(\alpha_i) + \sum_{j=1}^{N} m_j E(\beta_j) + \sum_{i,j=1}^{N} l_{ij} E(\alpha_i, \beta_j) \leq I_L, \tag{3.3.6}$$

where α_i and β_i are the angles of the measurements in Alice and Bob site respectively, n_i, m_j and l_{ij} are coefficients defining the inequality and I_L is the local bound. The inequality I can be represented schematically as a table

$$I = \left(\begin{array}{c|cccc} & m_1 & m_2 & \dots & m_N \\ \hline n_1 & l_{11} & l_{12} & \dots & l_{1N} \\ n_2 & l_{21} & l_{22} & \dots & l_{2N} \\ \dots & \dots & \dots & \dots & \dots \\ n_N & l_{N1} & l_{N2} & \dots & l_{NN} \end{array} \right) \leq I_L. \tag{3.3.7}$$

Note that this table has the same form as the one introduced in [13], but here coefficients correspond to expectations values and not to probabilities.

It is clear from Eq. (3.3.6) that, for a given quantum state, the term I is a function of the measurement angles on Alice's ($\{\alpha_1, \ldots, \alpha_N\}$) and Bob's site ($\{\beta_1, \ldots, \beta_N\}$), that is

$$I = I(\alpha_1, \ldots, \alpha_N, \beta_1, \ldots, \beta_N). \tag{3.3.8}$$

This function is not linear in terms of its $2N$ variables, but numerical optimizations can be used in order to find the optimal measurement settings for Alice and Bob, which will provide the largest possible value of I for the state under consideration.

For the nonlocality tests reported in this paper, we perform a partial tomography of the polarization entangled states that we prepare, as explained in the previous section. This allows us to reconstruct the state in the plane of the Bloch sphere corresponding to linear polarizations. Then, we numerically find the optimal measurement settings for this state, restricting them to lie on the xz plane. These settings are used for the experimental violation of the Bell inequalities taken in consideration.

3.4 Nonlocality Tests with Multiple Settings

3.4.1 The Entanglement Source

We use the commercial entanglement source sold by the QuTools company [11], to which we brought some minor modifications. This source generates photon pairs entangled in polarization from spontaneous parametric down conversion (SPDC) in a bulk BBO crystal. The nonlinear crystal is cut for a Type II phase-matching and is pumped by a continuous wave diode laser at 405 nm. Photon pairs at 810 nm are generated, filtered and collected into single mode fibers. Linear polarizers at the Alice and Bob site can measure polarizations respectively at the angles α and β with respect to the vertical direction, as shown by the sketch in Fig. 3.2. Photons are then detected using a silicon avalanche photodiodes (Si-APDs) with efficiencies at the wavelength of the photons of 48 and 55 % respectively. The coincidences are measured by a time-to-digital converter (TDC) and a coincidence rate of 4.2 kHz is

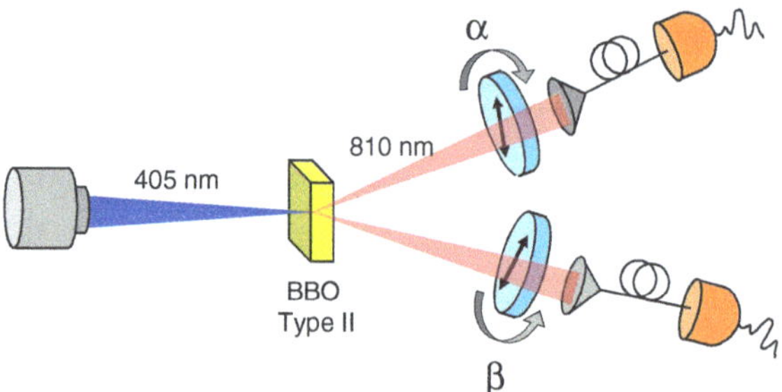

Fig. 3.2 Simple sketch of the polarization entanglement source. Linear polarizers at the Alice and Bob site are used for the settings necessary for the measurement of the Bell inequalities

measured for 15 mW of input pump power in the absence of the linear polarizers. Other details on this source can be found in [21].

For a pair of polarization directions measured by two polarizers, the number of coincidences in a time interval of 20 s is measured. The error associated to the coincidence rate is estimated according to poissonian statistics. The inequalities that we want to test require to measure a large number of correlations. One correlation term is given by measuring one photon in one basis and the other photon in another one. Measuring one photon in one basis means projecting it into two orthogonal polarization directions. So the measurement of one correlation term require 4 different settings for the two polarizers.

3.4.2 The Characterization of the State

We perform a partial tomography of the state generated by the source on the plane of the Bloch sphere corresponding to the linear polarizations. We determine the 8 correlation values indicated in the Sect. 3.3.1, demanding to measure on each side the operators in polarization:

$$\begin{aligned}\sigma_z &\equiv |H\rangle\langle H| - |V\rangle\langle V|,\\ \sigma_x &\equiv |+\rangle\langle +| - |-\rangle\langle -|,\\ \mathbb{1} &\equiv |H\rangle\langle H| + |V\rangle\langle V| \equiv |+\rangle\langle +| + |-\rangle\langle -|,\end{aligned}$$

where $|H\rangle$ and $|V\rangle$ correspond to the states of horizontal and vertical polarizations respectively and $|+(-)\rangle = \frac{1}{\sqrt{2}}(|H\rangle \pm |V\rangle)$. In the Table 3.1 a list of the measured expectation terms for the partial tomography of the prepared state is given.

The error associated to a correlation term is obtained simply by statistical propagation of the error in the coincidences on which it depends. The marginal terms, the last four in the table, have been measured twice using two different bases on Bob's site. The difference between these two values reflects in errors which are larger with

Table 3.1 List of the expectation terms for the partial tomography of the prepared entangled state

Expectation term	Exp	Th
$\langle\sigma_z \otimes \sigma_z\rangle$	-0.9649 ± 0.0012	-1
$\langle\sigma_x \otimes \sigma_x\rangle$	-0.9344 ± 0.0017	-1
$\langle\sigma_z \otimes \sigma_x\rangle$	0.1053 ± 0.0045	0
$\langle\sigma_x \otimes \sigma_z\rangle$	-0.0201 ± 0.0048	0
$\langle\sigma_z \otimes \mathbb{1}\rangle$	0.065 ± 0.034	0
$\langle\sigma_x \otimes \mathbb{1}\rangle$	0.036 ± 0.014	0
$\langle\mathbb{1} \otimes \sigma_z\rangle$	-0.078 ± 0.020	0
$\langle\mathbb{1} \otimes \sigma_x\rangle$	-0.015 ± 0.019	0

The measured values (Exp) can be compared to the theoretical ones (Th) that are expected from a perfect singlet state respect to the other expectation terms. Note that errors on the measured expectation terms need to be considered for the partial tomography in Table 3.1 to define density matrix which is definite positive. In some cases, in order to avoid the negativity of the density matrix attainable from a tomography, some techniques, such as the maximum likelihood estimation, need to be employed [19].

It is evident that the singlet state prepared with the commercial source has some imperfections. There is a mixture component in the state since the expectation values are slightly different to the theoretical ones. The state is also unbalanced in the two orthogonal bases $\{|H\rangle, |V\rangle\}$ and $\{|+\rangle, |-\rangle\}$, which confirms that it is not maximally entangled.

The problem of optimization of the measurement settings for the nonlocality tests is necessarily limited to the plane on which we have limited the tomography. We measure 9 inequalities requiring in total 332 settings of the linear polarizers in the Alice and Bob site. These optimal settings have in some cases differences of some degrees with respect to the standard settings used for obtaining a maximum violation with a perfect singlet state.

3.4.3 CHSH Inequality

First of all, in order to check the validity of our optimization method, we measure the CHSH inequality using the standard settings for the singlet state, then the same inequality with the settings optimized for the prepared state. According to the notation given in Sect. 3.3.2, we can represent the CHSH inequality by the following table

$$I_{CHSH} = \left(\begin{array}{c|cc} & 0 & 0 \\ \hline 0 & 1 & 1 \\ 0 & -1 & 1 \end{array}\right) \leq 2. \quad (3.4.1)$$

For the CHSH test with optimal settings we obtain a value of 2.731 ± 0.015, clearly enhanced with respect to 2.691 ± 0.015, obtained by adopting the standard settings. These two values correspond to violations of the local bound of respectively 49 and 46 standard deviations. This confirms the validity of the settings' optimization. Actually, we expected from the partial tomography a value for CHSH of 2.683 with optimal settings and 2.662 with the standard ones. This little discrepancy could be explained by small variations of the prepared state between the tomography and the final measurements.

3.4.4 Inequalities Inequivalent to CHSH

We then measure Bell inequalities inequivalent to CHSH, in particular I_{3322} [13] and two different types of I_{4422}, called AS_1 and AS_2 [14, 22]. These inequalities are facets of their corresponding Bell polytope [20]. They are thus optimal to detect nonlocality for some correlations in scenarios involving three and four settings. Note that I_{3322} asks to measure not only joint correlations, but also four marginal probabilities. On the contrary, the two I_{4422} inequalities that we want to measure are the only two of this kind to require correlation terms uniquely. This makes their measurement simpler from a conceptual and experimental point of view. In the following, the coefficients of these inequalities are given

$$I_{3322} = \left(\begin{array}{c|ccc} & 1 & 1 & 0 \\ \hline 1 & -1 & -1 & -1 \\ 1 & -1 & -1 & 1 \\ 0 & -1 & 1 & 0 \end{array}\right) \leq 4, \tag{3.4.2}$$

$$AS_1 = \left(\begin{array}{c|cccc} & 0 & 0 & 0 & 0 \\ \hline 0 & 1 & 1 & 1 & 1 \\ 0 & 1 & 1 & 1 & -1 \\ 0 & 1 & 1 & -2 & 0 \\ 0 & 1 & -1 & 0 & 0 \end{array}\right) \leq 6, \tag{3.4.3}$$

$$AS_2 = \left(\begin{array}{c|cccc} & 0 & 0 & 0 & 0 \\ \hline 0 & 2 & 2 & 1 & 1 \\ 0 & 2 & -1 & -1 & -2 \\ 0 & 1 & -1 & -2 & 2 \\ 0 & 1 & -2 & 2 & 1 \end{array}\right) \leq 10. \tag{3.4.4}$$

Note that this notation is different with respect to that used in [14]. We call I^{exp} the experimental value of the Bell parameter. For each of the three different inequalities, we observe a violation of the local bound I_L (Table 3.2). In the Table 3.2 the result for CHSH is that with optimal settings. The violations for the two I_{4422} inequalities are stronger than for I_{3322}. We evaluate this aspect by considering the resistance to noise of the three different inequalities. If some white noise were added with

Table 3.2 Measurement of the CHSH inequality and of the inequalities inequivalent to CHSH

	I_L	I^{exp}	I^{tom}	$I^{exp} - I_L$ (σ units)	$p_{\text{noise}}(\%)$
I_{CHSH}	2	2.731 ± 0.015	2.683	49	27
I_{3322}	4	4.592 ± 0.024	4.769	25	13
AS_1	6	7.747 ± 0.026	7.750	67	23
AS_2	10	12.85 ± 0.030	12.819	95	22

I_L is the local bound, I^{exp} is the value of the Bell parameter obtained experimentally with the optimized settings, I^{tom} is the expected value from the partial tomography, $I^{exp} - I_L$ is the difference between the obtained value and the local bound in terms of number of standard deviations σ and $p_{\text{noise}}(\%)$ is the critical level of white noise that can be added to the system while still keeping a violation probability p_{noise} to our state ρ, we would obtain a state with a density matrix $\rho_{\text{noise}} = p_{\text{noise}} \frac{\mathbb{1}}{4} + (1 - p_{\text{noise}})\rho$. Now, it is easy to show that the violation would not be observed anymore if $p_{\text{noise}} > 1 - \frac{I_L}{I^{exp}}$. For I_{3322} adding 13 % of white noise to the state compromises the violation. For the other inequalities the tolerance to the noise is higher and the CHSH inequality confirms to be the most robust to noise.

In all the cases the obtained violations agree quite well with what we expected with the tomography.

3.4.5 Chained Inequalities

Finally, we measure chained inequalities with a number of settings per side from 2 to 6. The chained inequality with N settings for Alice ($\{\alpha_1, \alpha_2, \ldots, \alpha_N\}$) and N for Bob ($\{\beta_1, \beta_2, \ldots, \beta_N\}$) can be written as a correlation inequality as

$$\begin{aligned} I_{chain}^{(N)} &= E(\alpha_1, \beta_1) + E(\beta_1, \alpha_2) + E(\alpha_2, \beta_2) + \cdots \\ &+ E(\alpha_N, \beta_N) - E(\beta_N, \alpha_1) \leq 2(N-1). \end{aligned} \tag{3.4.5}$$

Contrary to the previous inequalities, only the chained inequality with $N = 2$, which is equivalent to CHSH, is a facet of the Bell polytope.

Since the chained inequalities require $2N$ correlation terms, the number of settings required for their measurement scales linearly with N. We thus limit ourselves to 6 settings per side. In the Table 3.3 a list of the measured inequalities is given. For $N = 2$ we have the result for the CHSH inequality with optimal settings. All the inequalities are violated in a way consistent with the expected results and with a large number of standard deviations of difference with respect to the local bound. However, it is interesting to note that the larger the number of settings, the weaker the violation. Indeed, the fraction of noise that we can add to the system and still keep the violations decreases for an increasing number of settings.

Table 3.3 Measurement of the chained inequalities with N settings per side

N	I_L	I^{exp}	I^{tom}	$I^{exp} - I_L$ (σ units)	p_{noise} (%)
2	2	2.731 ± 0.015	2.683	49	27
3	4	4.907 ± 0.019	4.925	48	18
4	6	7.018 ± 0.023	6.999	44	15
5	8	8.969 ± 0.026	8.996	37	11
6	10	10.91 ± 0.028	10.954	33	8

3.5 Application of the Chained Inequalities

As recalled in the Sect. 3.2, the violation of the chained inequalities can be linked with numerous problems. In the following we briefly mention what the violations we observed allow us to conclude with respect to the EPR2 model of nonlocality and to true randomness.

3.5.1 EPR2 Nonlocality

Chained inequalities can be used to put an upper bound on the local content of the prepared state according to the EPR2 approach [16]. Indeed, as already explained in the Sect. 3.2, one can imagine that only part of the photon pairs produced by the source has nonlocal properties, while the other ones behave in a purely local way. In such a situation, the measured value of $I^{(N)}_{chain}$ decomposes as a sum of the local bound I_L with probability p_L and of some possibly larger value I with probability $1 - p_L$. The latter is only bounded by the no-signaling (NS) principle, i.e. the impossibility of faster-than-light communication. Therefore

$$I^{exp} = p_L I_L^{(N)} + (1 - p_L) I_{NS}^{(N)}. \tag{3.5.1}$$

Since the local and the no-signaling values of the I quantity are at most $I_L^{(N)} = 2(N-2)$ and $I_{NS}^{(N)} = 2N$ respectively, we have the following bound on the local part of the produced state:

$$p_L \leq N - \frac{I^{exp}}{2}. \tag{3.5.2}$$

To illustrate this bound, let us see how it applies to noisy singlet states, i.e. Werner states $\rho_W = V|\psi^-\rangle\langle\psi^-| + (1 - V)\frac{\mathbb{1}}{4}$. For these states the maximum value of $I^{(N)}$ is $2NV\cos(\frac{\pi}{2N})$.Thus, the local probability p_L^W associated to a Werner state ρ_W depends on the visibility V of the state and it is given by $p_L^W = N(1 - V\cos(\frac{\pi}{2N}))$. The behaviour of this function is shown in the Fig. 3.3 for different values of the visibility V.

For the pure singlet state ($V = 1$), p_L^W tends to 0 for an infinite number of settings, confirming the fact that the singlet state is fully nonlocal [17]. On the contrary, when V is smaller than 1, p_L^W decreases until a certain limit value. Therefore, the number of settings which are useful for lowering the upper bound on the local content of the state is limited and changes according to the visibility V.

We observe a similar effect for the state produced by our source. In Fig. 3.4 we represent the violations of the measured chained inequalities in terms of local probabilities, as calculated from 3.5.2. The red line corresponds to the expected values for p_L obtained by the knowledge of the state. Our state is more complex than a Werner state but it can be approximated by a Werner state with visibility $V = 0.94$.

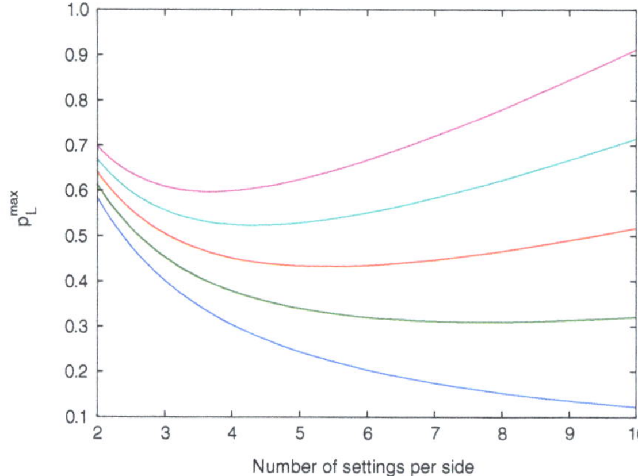

Fig. 3.3 Largest local part associated to a Werner state for different values of $V = 1$, 0.98, 0.96, 0.94, 0.92 (from the *bottom* to the *top*)

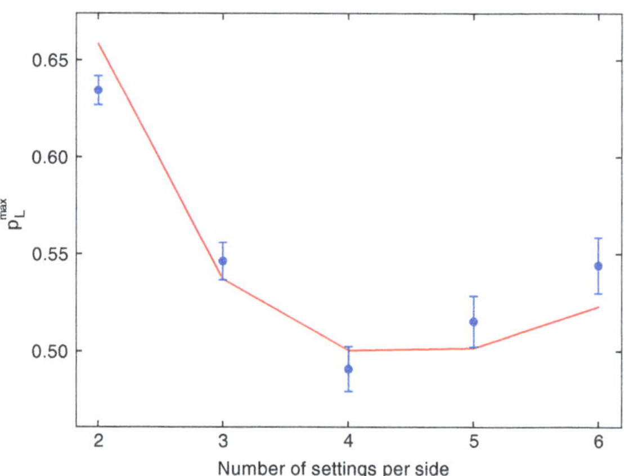

Fig. 3.4 Largest local part as a function of the number of settings per side. The *red line* corresponds to the expected values for our state

The minimum value of p_L that we measure is 0.491 ± 0.012 with a chained inequality of 4 settings per side. This result means that at least half of the photon pairs produced by the source are nonlocal, a result that cannot be shown with the CHSH inequality. To our knowledge, this is the first experiment that fixes an upper bound on the local part of the quantum correlations according to the EPR2 approach [23].[2] Similar work in progress is expected to show a lower value of this quantity [24].

3.5.2 Randomness Certified by the No-signaling Principle

When measuring a singlet state, locally random outcomes can be observed. It has been recently shown that violation of a Bell inequality can certify that this randomness truly emerges during the experiment, in the sense that no algorithm can possibly

[2] Note that existing experimental results could be reinterpreted to provide such a bound as well.

predict the measured outcomes [25]. Indeed, if such algorithm existed prior to the experiment, it could be considered as a local hidden variable, and no violation of a Bell inequality can be observed with only local hidden variables.

Note that in order to discard any such algorithm, the detection loophole should be closed during the experiment. Indeed, such an algorithm could in principle inform detectors when they should click according to some detection loophole model [26]. In order to avoid this issue, we assume fair sampling of the detected events: The state of the particles coming onto a detector does not affect the fact that a detector fires or not, so that the detected pairs of particles fairly represent the ones produced by the source.

In order to quantify the amount of true randomness that can be found in some experimental results, one must consider all the marginal probabilities $P(a|x)$ that Alice finds the outcome a when she measures x, which are compatible with the observed Bell inequality violation I^{exp}. We searched for the largest marginal probability $P(a|x)$ numerically among all possible quantum correlations, i.e. correlations that can be achieved by measuring a quantum state, as well as among all no-signaling correlations, denoting this largest quantity by $P^*(A|X)$.

The computed upper bounds $P^*(A|X)$ for the chained inequalities with up to 6 settings per party are shown in Fig. 3.5 together with our experimental results. The lowest bound on the marginal probability that we can certify here is $P^*(A|X) = 0.684 \pm 0.014$, achieved for the CHSH inequality. However, if we consider the bound imposed by the no-signaling principle only, then the strongest one is $P^*(A|X) = 0.7455 \pm 0.0057$, achieved by the chained inequality with $N = 4$ settings per party.

Note that in order to extract a truly random bit string out of measured outcomes, classical key distillation techniques should be used [27]. The ratio between the number of measured bits and the number of truly random, uniformly distributed, bits that is produced by this procedure is given by the min-entropy of Alice's outcome A conditioned on her measurement choice X: $H_{\min}(A|X) = -\log_2 \max_a P^*(A|X)$. In our case we find $H_{\min} = 0.55 \pm 0.03$ for the CHSH violation, meaning that approximately one random bit every two measurements have been created.

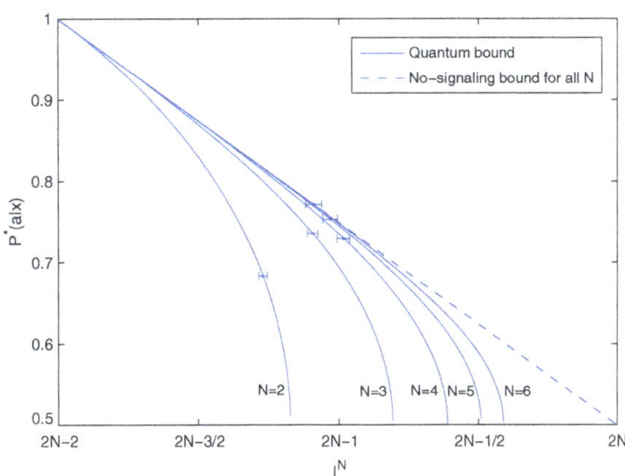

Fig. 3.5 Upper-bound on the marginal probability distribution, as a function of the inequalities violation. The bound implied by the no-signaling principle is identical for all chained inequalities

3.6 Conclusions

We have shown that non trivial nonlocality tests alternative to CHSH can be performed even with a rather simple commercial source. In particular, we have measured Bell inequalities with multiple measurement settings. In particular, we have reported violations of I_{3322}, of two I_{4422} and of chained inequalities with up to 6 settings per side. Violations of these inequalities have not been shown before. Moreover, by using the chained inequalities, we have put an upper bound on the local content of the prepared state at 0.491 ± 0.012, meaning that at least half of the photon pairs produced by the source have nonlocal correlations. We have also quantified the amount of true randomness created in the experiment. Therefore, even with an extremely simple setup, it is possible to implement not trivial nonlocality tests and get interesting conclusions on the nonlocal properties of the source and on the randomness produced in the experiment. It emphasizes the richness of nonlocality and the importance of the no-signaling and of the detection loophole. This could be of interest for undergraduate laboratory courses.

Acknowledgments This work is supported by Qessence and NCCR-QP. We would like to thank Antonio Acin, Denis Rosset and Y.-C. Liang for valuable discussions, suggestions and remarks.

References

1. A. Einstein, B. Podolsky, N. Rosen, Can Quantum-Mechanical Description of Physical Reality Be Considered Complete? Phys. Rev. **47**, 777 (1935)
2. J.S. Bell, On the Einstein-Podolski-Rosen paradox. Phys. **1**, 195 (1964)
3. J.F. Clauser, M.A. Horne, A. Shimony, R.A. Holt, Proposed Experiment to Test Local Hidden-Variable Theories. Phys. Rev. Lett. **23**, 880 (1969)
4. A. Aspect, P. Grangier, G. Roger, Experimental Tests of Realistic Local Theories via Bell's Theorem. Phys. Rev. Lett. **47**, 460 (1981)
5. A. Aspect, J. Dalibard, G. Roger, Phys. Rev. Lett. **49**, 1804 (1982)
6. G. Weihs, T. Jennewein, C. Simon, H. Weinfurter, A. Zeilinger, Phys. Rev. Lett. **81**, 5039 (1998)
7. W. Tittel, J. Brendel, N. Gisin, H. Zbinden, Phys. Rev. A **59**, 4150 (1999)
8. M.A. Rowe, D. Kielpinski, V. Meyer, C.A. Sackett, W.M. Itano, C. Monroe, D.J. Wineland, Nature (London) **409**, 791 (2001)
9. D.N. Matsukevich, P. Maunz, D.L. Moehring, S. Olmschenk, C. Monroe, Phys. Rev. Lett. **100**, 150404 (2008)
10. T. Vértesi, S. Pironio, N. Brunner, Closing the Detection Loophole in Bell Experiments Using Qudits. Phys. Rev. Lett. **104**, 060401 (2010)
11. See http://www.qutools.com
12. J.B. Altepeter, E.R. Jeffrey, P.G. Kwiat, S. Tanzilli, N. Gisin, A. Acín, Experimental Methods for Detecting Entanglement. Phys. Rev. Lett. **95**, 033601 (2005)
13. D. Collins, and N. Gisin, "A relevant two qubit Bell inequality inequivalent to the CHSH inequality", Journ. of Phys. A: Math. and Gen. 37, 1775 (2004).
14. N. Brunner, N. Gisin, Partial list of bipartite Bell inequalities with four binary settings. Phys. Lett. A **372**, 3162 (2008)
15. S. Braunstein, C. Caves, Wringing out better Bell inequalities. Annals of Phys. **202**, 22 (1990)
16. A. Elitzur, S. Popescu, D. Rohrlich, Quantum nonlocality for each pair in an ensemble. Phys. Lett. A **162**, 25 (1992)

17. J. Barrett, A. Kent, S. Pironio, Maximally Nonlocal and Monogamous Quantum Correlations. Phys. Rev. Lett. **97**, 17 (2006)
18. V. Scarani, Local and nonlocal content of bipartite qubit and qutrit correlations. Phys. Rev. A **77**, 042112 (2008)
19. D.F.V. James, P.G. Kwiat, W.J. Munro, A.G. White, Measurement of qubits. Phys. Rev. A **64**, 052312 (2001)
20. R. F. Werner, and M. M. Wolf, "Bell inequalities and Entanglement", Quantum Inf. Comput. 1:3, 1–25.
21. P. Trojek, C. Schmid, M. Bourennane, H. Weinfurter, Ch. Kurtsiefer, Compact source of polarization-entangled photon pairs. Opt. Expr. **12**, 276 (2004)
22. D. Avis, H. Imai, and T. Ito, On the relationship between convex bodies related to correlation experiments with dichotomic observables", Journ. of Phys. A: Math. and Gen. 39, 11283 (2006)
23. Note that existing experimental results could be reinterpreted to provide such a bound as well
24. L. Aolita et al. (unpublished)
25. S. Pironio, A. Acín, S. Massar, A. Boyer de la Giroday, D.N. Matsukevich, P. Maunz, S. Olmschenk, D. Hayes, L. Luo, T.A. Manning, C. Monroe, Random numbers certified by Bell's theorem. Nat. **464**, 1021 (2010)
26. N. Gisin, B. Gisin, Phys. Lett. A **297**, 279 (2002)
27. S. Ronen, Recent developments in extractors. Bulletin of the European Association for Theoretical Computer Science **77**, 67 (2002)

Chapter 4
Nonlocality with Three and More Parties

In the precedent chapters, we focused on Bell-type experiments involving two parties only. While this is the simplest case, and indeed the most often discussed one, the idea of local correlations can be extended straightforwardly to multipartite scenarios involving more parties.

Labeling the (output,input) of a third party Charly by (c,z), the locality condition (1.1.1) generalises to:

$$P(abc|xyz) = \sum_{\lambda} p(\lambda) P_A(a|x,\lambda) P_B(b|y,\lambda) P_C(c|z,\lambda), \tag{4.0.1}$$

and similarly for more parties. Tripartite correlations $P(abc|xyz)$ are then referred to as nonlocal if and only if they cannot be decomposed as (4.0.1).

4.1 Defining Genuine Multipartite Nonlocality

Just like entanglement can have more forms in a multipartite scenario than in the bipartite case [1], it is easy to realize that the definition (4.0.1) does not capture the whole potential of nonlocality in a tripartite scenario. Consider indeed some bipartite nonlocal correlations $P_{AB}(ab|xy)$ and arbitrary statistics for Charly $P_C(c|z)$. The product of the two distributions $P(abc|xyz) = P_{AB}(ab|xy)P_C(c|z)$ violates (4.0.1) and is thus nonlocal. However it is clear that Charly plays no role in the nonlocality of these correlations. These correlations are thus not genuinely three-way nonlocal.

This observation was first made by Svetlichny in 1987 [2], who proposed an inequality capable of certifying (if violated) that correlations cannot be explained by a mechanism involving fewer than 3 parties. This is the Svetlichny inequality

$$S = E_{111} + E_{112} + E_{121} - E_{122} + E_{211} - E_{212} - E_{221} - E_{222} \leq 4 \tag{4.1.1}$$

J.-D. Bancal, *On the Device-Independent Approach to Quantum Physics*, Springer Theses,
DOI: 10.1007/978-3-319-01183-7_4,

with $E_{xyz} = \sum_{a,b,c=0}^{1}(-1)^{a+b+c}P(abc|xyz)$, which is satisfied by all tripartite correlations of the form

$$P(abc|xyz) = \sum_{\lambda} p_1(\lambda)P_{AB}(ab|xy,\lambda)P_C(c|z,\lambda) + \sum_{\lambda} p_2(\lambda)P_{AC}(ac|xz,\lambda)P_B(b|y,\lambda) + \sum_{\lambda} p_3(\lambda)P_{BC}(bc|yz,\lambda)P_A(a|x,\lambda) \tag{4.1.2}$$

with $p_i \geq 0$ and $\sum_{\lambda} p_1(\lambda) + p_2(\lambda) + p_3(\lambda) = 1$.

If being unable to decompose some tripartite correlations $P(abc|xyz)$ in the form of (4.1.2) is sufficient to conclude that none of the parties was separated from the other ones in the process that created these correlations, it was pointed out recently that this condition is not always necessary (see [3, 4]).

To understand why this is the case, let us consider the situation in which the three measurement events producing a, b, and c, are not simultaneous but follow an order: Alice measures first, then Bob, and finally Charly ($A < B < C$). If decompositions of the form (4.1.2) exist for the observed correlations we might want to conclude that these correlations can be reproduced by some interaction between pairs of parties. Yet, this is not possible if every such decomposition happens to contain P_{AB} terms that are signalling from B to A, i.e. such that $\sum_b P_{AB}(ab|xy,\lambda)$ depends on y. Indeed, in the considered configuration ($A < B < C$), y can always be chosen freely after a. The distribution of a thus cannot depend on y.

It thus seems important, from a physical point of view, to consider decompositions (4.1.2) that are compatible with the situation in which the correlations are produced. In order to conclude something about the nature of correlations that is independent of the situation in which they appear, we suggest to require a consistent decomposition (4.1.2) to exist for all possible measurement situations. Thus, we say that correlations are *Svetlichny-sequential* if they can be decomposed as

$$P(abc|xyz) = \sum_{\lambda} p_1(\lambda)P^{T_{AB}}(ab|xy,\lambda)P_C(c|z,\lambda) + \sum_{\lambda} p_2(\lambda)P^{T_{AC}}(ac|xz,\lambda)P_B(b|y,\lambda) + \sum_{\lambda} p_3(\lambda)P^{T_{BC}}(bc|yz,\lambda)P_A(a|x,\lambda) \tag{4.1.3}$$

for every possible ordering of the measurements. Non-Svetlichny-sequential correlations are then called *genuinely tripartite nonlocal*. For correlations that are not genuinely tripartite nonlocal in this sense, a biseparable model cannot be constructed coherently for all possible ordering of the measurements. Here $P^{T_{AB}}(ab|xy,\lambda)$ depends on the order of measurement between Alice and Bob. Namely, $P^{T_{AB}}(ab|xy,\lambda) = P^{A<B}(ab|xy,\lambda) = P_A(a|x,\lambda)P_B(b|y,ax\lambda)$ if Alice measures first, and

$P^{T_{AB}}(ab|xy,\lambda) = P^{B<A}(ab|xy,\lambda) = P_B(b|y,\lambda)P_A(a|x,by\lambda)$ if Bob measures first.

Note that the problem we just mentioned can also be avoided by requiring all terms appearing in (4.1.2) to be no-signalling. Namely, if all bipartite correlations terms $P_{AB}(ab|xy)$, $P_{AC}(ac|xz)$ and $P_{BC}(bc|yz)$ satisfy the no-signalling constraints (1.1.1) in Eq. (4.1.2), then it is always possible to reproduce these correlations with communication between only two parties, independently of the order in which the parties perform their measurements. We call correlations having such a decomposition *Svetlichny-no-signalling*. One can show that the condition obtained by requiring terms in (4.1.2) to be no-signalling is strictly stronger than the one given by Eq. (4.3.1) (which is also strictly stronger than (4.1.2) without the no-signalling requirement). Some correlations can thus be reproduced with communication between two parties only, and in a way that is consistent with all possible orders of the measurement, even though they are not Svetlichny-no-signalling (see [3] for more information about that).

To conclude, note that the three definitions we just discussed generalize straightforwardly to scenarios involving an arbitrary number of parties. Moreover, they all reduce to the locality condition (1.1.1) in the bipartite case. This is possibly one reason why details about the time ordering of the measurement events is usually not discussed when talking about nonlocal correlations. Yet, a precise account of the order in which measurements are performed is crucial in several situations, as shown here, and in the last two chapters of this thesis.

4.2 Multipartite Bell-Like Inequalities

If the CHSH inequality (4.2.2) found many applications, it is certainly because of its outstanding properties , but probably also because of its simplicity. Indeed, the CHSH inequality has a lot of symmetries: it is for instance invariant under permutation of parties, so that no party plays a special role, and it only involves correlations between the parties' outcomes, so that the specific outcomes of a party play no specific role independently of the other party's outcome.

Since the structure of the local polytope quickly gets too complicated to allow a direct computation of its facets [5] when more than two parties, inputs or outputs are considered, it seems natural, when considering such scenarios, to first restrict one's attention to Bell inequalities having properties similar to the ones we just mentioned. This allows one to simplify the analysis while leaving a hope that the results found in this way can be useful, since the CHSH inequality satisfies these constraints.

In this perspective, we note that a complete description of the Bell inequalities which involve only full-correlations terms could be found for all $(n, 2, 2)$ scenarios in [6, 7]. Here we denote by (n, m, k) the Bell scenario involving n parties, each with m possible measurement settings producing one out of k possible outcomes. On the other hand, we describe in [8] how the search for Bell inequalities that are symmetric under permutations of the parties can be simplified by considering projections of the local polytope, which are much easier to solve than the full polytopes. This allows us in turn to discover many families of Bell and Svetlichny inequalities for several scenarios (c.f. [8] for more details).

Here, we consider a special form of Bell expressions that is both symmetric under permutation of the parties and only involves full-correlations. We show that many inequalities presented throughout the years in the literature have this form. This allows us to propose a natural generalization of them to general (n, m, k) scenarios. Moreover, we show that several bounds on these expressions can be easily computed once some bound for the corresponding few-party inequality is known.

4.2.1 A General Structure for (n, m, k) Scenarios

Let us consider the (n, m, k) Bell scenario. Denoting by $\vec{s} = (s_1, \ldots, s_n) \in \{0, 1, \ldots, m-1\}^n$ the settings of all parties, and $\vec{r} = (r_1, \ldots, r_n) \in \{0, 1, \ldots, k-1\}^n$ their results, we write the following Bell expression:

$$\Omega_{n,m,k;f} = \sum_{\vec{s}} \sum_{\vec{r}} f\left([\mathbf{s}]_m, \left[\mathbf{r} - \left\lfloor \frac{\mathbf{s}}{m} \right\rfloor \right]_k \right) P(\vec{r}|\vec{s}) \qquad (4.2.1)$$

where $\mathbf{s} = \sum_{i=1}^{n} s_i$ and $\mathbf{r} = \sum_{i=1}^{n} r_i$ are the sums of all parties' inputs and outputs, $\lfloor x \rfloor$ is the integer part of x, $[x]_y = x - y\lfloor x/y \rfloor$ the modulo function, and $f : \{0, \ldots, m-1\} \times \{0, \ldots, k-1\} \rightarrow \mathbb{R}$ is a real-valued function (defined by $m \times k$ real parameters) that fully characterizes $\Omega_{n,m,k;f}$. Clearly, this Bell polynomial is symmetric under permutation of the parties and only involves correlation terms, since only sums of all the parties' settings and outcomes matter.

A choice of function $f(s, r)$ allows one to write an expression for any scenario with n parties, m inputs and k outputs. Table 4.1 shows how different choices of this function allow to recover several Bell-like expressions used in the literature. In particular, the choice

$$f(s, r) = \delta_{s,0}\, r + \delta_{s,1}[-r]_k \qquad (4.2.2)$$

Table 4.1 A summary of some known Bell expressions that can be recovered as special cases of $\Omega_{n,m,k;f}$

n	m	k	$f(s, r)$	Bell expression
≥ 3, odd	2	2	$\delta_{s,0}\, r$	MABK [21–25]
≥ 3	≥ 2	2	$\delta_{s,0}\, r$	DIEW [26]
≥ 3	≥ 2	2	$\cos(\frac{\Delta - s}{m}\pi)\, r,\ \Delta \in \mathbb{R}$	DIEW [26]
≥ 2	≥ 2	≥ 2	$\delta_{s,0}\, r + \delta_{s,1}[-r]_k$	c.f. Fig. 4.1

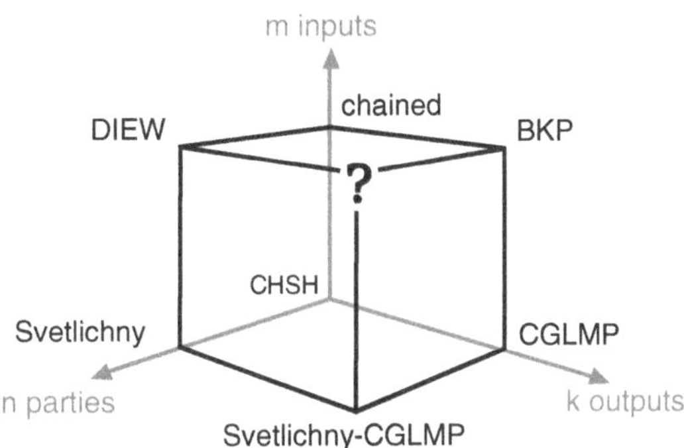

Fig. 4.1 Previously known families of Bell expressions (see [9–17] and [18, 19]) are recovered by Eq. (4.2.1) with the choice of parameter $f(s,r)=\delta_{s,0}\,r+\delta_{s,1}[-r]_k$. In particular, the CHSH expression is recovered for $n=m=k=2$. This provides a natural way to extend these inequalities to a general scenario (n,m,k)

allows one to generalize all the expressions represented in Fig. 4.1 to the (n,m,k) scenario. Note that the generalization obtained in this way was also discovered independently by [20].

4.2.2 Recursion Relation

By performing the change of variable $s_1'=[s_1+s_n]_m$, $r_1'=[r_1+r_n-\lfloor\frac{s_1+s_n}{m}\rfloor]_k$, Eq. (4.2.1) can be rewritten as

$$\Omega_{n,m,k;f}=\sum_{s_n=0}^{m-1}\sum_{r_n=0}^{k-1}\Omega_{n-1,m,k;f}^{(s_n,r_n)}\,P(r_n|s_n)\,, \qquad (4.2.3)$$

where $\Omega_{n-1,m,k;f}^{(s_n,r_n)}$ is equivalent upon permutation of inputs and outputs to an $(n-1)$-partite expression (4.2.1) conditioned on r_n and s_n. Since the $(n-1)$-partite polynomial is generated by the same function $f(s,r)$ as the n-partite one $\Omega_{n,m,k;f}$, this provides a way to relate the n-partite expression to polynomials of the same kind involving fewer parties. We describe below how this relation allows one to derive a number of bounds for $\Omega_{n,m,k;f}$.

4.2.2.1 Tsirelson Bounds

Given, a Tsirelson bound $\Omega_{2,m,k;f}\geq\beta_{2,m,k;f}^T$ on a bipartite Bell-like expression, Eq. (4.2.3) induces the following Tsirelson bound for the n-partite polynomial:

$$\Omega_{n,m,k;f}\geq m^{n-2}\,\beta_{2,m,k;f}^T\,. \qquad (4.2.4)$$

Nontrivial Tsirelson bounds can thus be easily deduced for these multipartite Bell inequalities, thanks to their special structure.

4.2.2.2 Generalized Svetlichny Bounds

By considering a scenario in which n parties are gathered into two groups, Svetlichny deduced an inequality which detects when interaction between all parties must have happened [2]. More generally, the amount of interaction needed between n parties in order to reproduce some correlations can be quantified by the maximal number of groups G into which the parties can be separated while still being able to reproduce these correlations (c.f. Fig. 4.2). Within each group, parties are allowed to communicate their inputs to each other, and to agree on which outputs they want to outcome, but no communication is allowed between the different groups.[1]

Thanks to relation (4.2.3), the bound of any $\Omega_{n,m,k;f}$ polynomial that can be achieved with parties distributed into G groups can be obtained from the local bound of the polynomial with $n = \text{G}$. Denoting this bound by $\beta^L_{\text{G},m,k;f}$ gives the following bound for $\Omega_{n,m,k;f}$ upon separation of the parties into G groups:

$$\Omega_{n,m,k;f} \geq m^{n-\text{G}}\,\beta^L_{\text{G},m,k;f}\,. \tag{4.2.5}$$

Note that similar bounds were already derived for correlations obtained by measuring quantum states that are positive under partial transposition across all partitions of the n systems into G subsystems [27].

Application to quantum states. Violation of (4.2.5) allows one to put an upper bound on the number of groups G into which the parties can be distributed in order for them to be able to reproduce the observed correlations with local operation and communication within the groups. Here, we consider the above bounds in the case $m = k = 2$. It turns out that in order to test for an even or odd number of groups it is useful to consider two different Bell polynomials (c.f. Chap. 5). For G odd, we thus choose $f(s,r) = \delta_{s,0}\,r$, and when G is even we choose $f(s,r) = \delta_{s,0}\,r + \delta_{s,1}[-r]_k$.

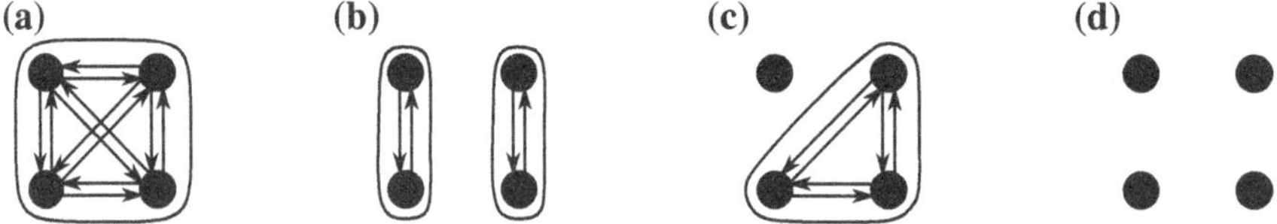

Fig. 4.2 In a grouping models, n parties can be shared out into G groups. Arbitrary communication is allowed between parties belonging to the same group, but no communication is allowed between different groups. **a** with $\text{G} = 1$, every no-signalling correlation can be reproduced by the model. **b** and **c** $\text{G} = 2$: Parties are shared into two groups. This corresponds to the usual Svetlichny model. Correlations that cannot be reproduced here are genuinely multipartite nonlocal. **d** For $\text{G} = n$, the model coincides with the usual local model. Any $2 < \text{G} < n$ allows one to interpolate between the local and usual Svetlichny models

[1] As discussed in Sect. 4.1, the order in which the different parties are measured should in principle be included in the model. However one can show that this order is not important here, i.e. the different definitions discussed in Sect. 4.1 coincide, because the inequalities we consider here only involve full-correlations.

Considering violations of these inequalities with measurements on the states in the GHZ family

$$|GHZ_\theta\rangle = \cos\theta|00\ldots0\rangle + \sin\theta|11\ldots1\rangle \tag{4.2.6}$$

shows that parties cannot be separated into more that $G = 1-2\lfloor\log_2\sin(2\theta)\rfloor$ groups in order to reproduce the achieved correlations (see Chap. 5).

By considering measurement on the n-partite W state

$$|W\rangle = \frac{1}{\sqrt{n}}(|10\ldots0\rangle + |01\ldots0\rangle + |00\ldots1\rangle), \tag{4.2.7}$$

we could also show that letting two parties interact is not enough to reproduce the correlations that can be observed as soon as $n \geq 3$. However, we found no violation for the bounds considered if the parties are separated into less than $n-1$ groups (see Chap. 5 for more details).

4.2.2.3 Biseparable Bounds

Finally, we note that Eq. (4.2.3) can also be used to deduce bounds on $\Omega_{n,m,k;f}$ that are satisfied upon measurement of biseparable quantum states. In particular, for $k=2$ outputs and $f(s,r) = g(s)\cdot r$, we show in [19] that the following biseparable bound holds:

$$\Omega_{n,m,2;g\cdot r} \geq \frac{1}{2}m^{n-2}\left(m\sum_{s=0}^{m-1}g(s) - \max_{j=0,\ldots,m-1}\left[\eta_j \csc\frac{\eta_j\pi}{2m}\left|\sum_{s=0}^{m-1}g(s)\,\omega_j^s\right|\right]\right) \tag{4.2.8}$$

where η_j is the greatest common divisor of $2j+1$ and m, and $\omega_j = \exp(i\pi(2j+1)/m)$.

The choice $g(s) = \delta_{s,0} + \delta_{s,1}$ allows one to write an entanglement witness that is well adapted to detect genuine multipartite entanglement in multipartite GHZ state. We use this witness in Chap. 6 of this thesis to detect genuine multipartite entanglement in a system of trapped ions.

4.3 Nonlocality from Local Marginals

In a multipartite system, every subset of parties constitutes a proper system in itself. The fact that these subsystems describe parts of the same total system requires them to satisfy some compatibility conditions: bipartite correlations $P(ab|xy)$ for instance are compatible with the tripartite ones $P(abc|xyz)$ if and only if $P(ab|xy) = \sum_c P(abc|xyz)$. While this condition is easy to check when the global correlations are known, checking whether several reduced states are compatible with each other is known as the marginal problem [28].

The marginal problem is trivial in the bipartite case: every single-party marginals $P(a|x)$ and $P(b|y)$ are always compatible with the joint distribution $P(ab|xy) = P(a|x)P(b|y)$. However, the situation is less evident in situations involving more parties. For instance, it is known that if P_{AB} violates the CHSH inequality, then it cannot be compatible with correlations P_{BC} violating CHSH as well; a phenomenon known as the monogamy of nonlocality [29].

Here we want to explore this multipartite constraint further. In particular, we ask whether some local marginals can witness nonlocality (or even genuine multipartite nonlocality), because the only global correlations with which they are compatible is (genuinely multipartite) nonlocal.

4.3.1 An Inequality

In order to answer this question in the tripartite case, we consider all bipartite correlations P_{AB}, P_{AC} and P_{BC}, which are compatible with Svetlichny-sequential correlations P_{ABC}. This set of correlations can be described as

$$\Pi = \{(P_{AB}, P_{AC}, P_{BC}) \mid \exists\ P_{ABC} \text{ s.t. 4.1.3 holds}\}. \tag{4.3.1}$$

Since Π is convex, it can be characterized with linear Bell-type inequalities.

Using the decomposition (4.3.1) and linear programming, one can show that the following inequality is satisfied by correlations in Π:

$$-\langle A_0(1+B_0+B_1+C_0)\rangle - \langle A_1(1+B_0+C_0+C_1)\rangle - \langle B_0+C_0+B_0C_0+B_1C_1\rangle \leq 4, \tag{4.3.2}$$

where $\langle A_x\rangle = P_A(0|x) - P_A(1|x)$, $\langle A_x B_y\rangle = P_{AB}(00|xy) - P_{AB}(01|xy) - P_{AB}(10|xy) + P_{AB}(11|xy)$ and similarly for the other parties. Violation of this inequality by some bipartite correlations thus implies that the only tripartite distributions with which they are compatible are genuinely-tripartite nonlocal.

Interestingly, this inequality can be violated by measuring a noisy W state

$$|W\rangle = p|W\rangle\langle W| + (1-p)\frac{\mathbb{1}}{8} \tag{4.3.3}$$

as soon as $p > 0.9548$, by using the measurement operators

$$\begin{aligned} A_0 &= \cos\alpha\sigma_z + \sin\alpha\sigma_x, & A_1 &= \cos\alpha\sigma_z - \sin\alpha\sigma_x \\ B_0 &= -\sigma_z, & B_1 &= \cos\beta\sigma_z + \sin\beta\sigma_x \\ C_0 &= -\sigma_z, & C_1 &= \cos\beta\sigma_z - \sin\beta\sigma_x \end{aligned} \tag{4.3.4}$$

with $\alpha = 3.6241$ and $\beta = 2.0221$.

However, no bipartite correlations achieved by measuring this state can violate a Bell inequality with binary inputs and outputs, because the bipartite reduced states

$$\rho_{red} = \frac{p}{3}(|01\rangle + |10\rangle)(\langle 01| + \langle 10|) + \frac{p}{3}|00\rangle\langle 00| + \frac{1-p}{4}\mathbb{1} \qquad (4.3.5)$$

satisfies the Horodecki criterion for every $0 < p < 1$ [30]. This shows that there exist bipartite quantum correlations that are local, but only compatible with genuinely tripartite-nonlocal correlations.

4.3.2 Conclusion

This illustrates the strength of the compatibility constraints that relates different parts of a system. Namely, the fact that subsystems are parts of the same system allows one to reveal properties of that system which were not apparent in its individual parts. Note that a similar result can be achieved with respect to entanglement (see [31] for more details).

4.4 Tripartite Nonlocal Boxes

Another way to get some insight into the structure of multipartite correlations is by studying the largest possible set of multipartite correlations. Within the no-signalling subspace, this is defined by all correlations satisfying the positivity conditions

$$P(abc|xyz) \geq 0 \; \forall \; a, b, c, x, y, z. \qquad (4.4.6)$$

The no-signalling polytope is thus easily characterized in terms of facets. Its extremal points, however, reveal that this polytope has a highly nontrivial geometric structure.

4.4.1 The Tripartite Nosignalling Polytope

Using the porta software [32], we could find all extremal points of the tripartite no-signalling polytope when all parties have binary inputs and outputs. After sorting these boxes into equivalence classes under permutation of inputs, outputs and parties, we could identify 46 families of extremal no-signalling boxes. A complete description of these families is given in [33].

Apart from the deterministic local strategies and the PR boxes, which are also extremal points of the bipartite nosignalling polytope, we found 44 families of truly tripartite boxes. Interestingly, three of them admit a decomposition of the form (4.3.1), and are thus not genuinely 3-way nonlocal in this sense.

Using the 46 classes of Bell inequalities that define the local polytope in this scenario [34], we studied the relation of each box with respect to each Bell inequality (c.f. [33]).

4.4.2 Conclusion

The reason why so many new no-signalling boxes appear when considering a scenario with more parties is still a mystery, which asks for further investigation. Nevertheless, the observation that the number of extremal boxes coincides with the number of Bell inequalities in this scenario was recently elucidated by Fritz [35]. It seems thus that a strong link exists between the set of nosignalling correlations and that of local correlations.

4.5 A Tight Limit on Quantum Nonlocality

As already mentioned in Sect. 1.1, quantum correlations can be nonlocal, but not as much as imposed by the no-signalling conditions (1.1.2) alone. Indeed, while no-signalling correlations can achieve a value of CHSH equal to 4 with a PR box, it is known since Tsirelson [36] that measurements on a quantum system cannot exceed $2\sqrt{2}$.

Despite recent advances on the subject [37, 38], the question why the nonlocality of quantum correlations is limited the way it is remains open. Here, we introduce a simple game for which quantum correlations provide no advantage, whereas general no-signalling correlations do. The inequality associated to this game is shown to identify part of the boundary separating quantum correlations from more general no-signalling correlations.

4.5.1 Can You Guess Your Neighbour's Input?

Let n parties form a circle as represented in Fig. 4.3. The game runs as follows: parties are given some input $x_i \in \{0, 1\}$, and are requested to output their right neighbour's input. The success that the parties have in playing this game can be measured with the following quantity:

$$\omega = \sum_{\mathbf{x}} q(\mathbf{x}) P(a_i = x_{i+1} \forall i | \mathbf{x}) \tag{4.5.1}$$

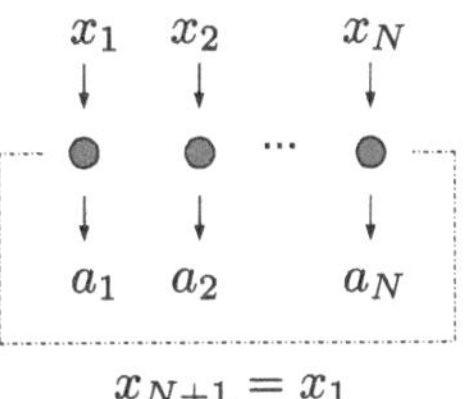

Fig. 4.3 Representation of the guess your neighbour's input (GYNI) nonlocal game. The goal is that each party outputs its right-neighbour's input, i.e. $a_i = x_{i+1}$

where $\mathbf{x} = (x_1, x_2, \ldots, x_n) \in \{0, 1\}^n$ denotes all the parties' inputs, $P(a_i = x_{i+1} \forall i|\mathbf{x})$ is the probability that the players obtain the correct outputs when they received the input string $\mathbf{x}$, and $q(\mathbf{x})$ is the probability that the set of inputs $\mathbf{x}$ is given to the parties.

One can show that the maximum value of ω that the parties can achieve by using classical or quantum resources is

$$\omega_c = \max_{\mathbf{x}}[q(\mathbf{x}) + q(\bar{\mathbf{x}})] \tag{4.5.2}$$

where $\bar{\mathbf{x}} = (\bar{x}_1, \bar{x}_2, \ldots, \bar{x}_n)$ with $\bar{x}_i = x_i \oplus 1$ is the negation of $\mathbf{x}$. Since this value is always achievable classically, it follows that the Bell inequality $\omega \leq \omega_c$ can never be violated quantum mechanically.

One might think that the reason why quantum correlations give no advantage in this game, as compared to local correlations, is that signalling is required in order to guess someone's input better than classically. But since quantum correlations are no-signalling, they cannot help it. However this is not totally true. Indeed, by choosing the distribution of inputs $q(\mathbf{x})$ adequately, it is sometimes possible to achieve a violation $\omega > \omega_c$ of the inequality without allowing the players to guess anything on other parties' inputs.

Consider for instance the following input distribution:

$$q(\mathbf{x}) = \begin{cases} 2^{1-\hat{n}} & \text{if } x_1 \oplus \ldots \oplus x_{\hat{n}} = 0 \\ 0 & \text{otherwise} \end{cases} \tag{4.5.3}$$

with $\hat{n}$ an odd number between 3 and n. In the case $n = \hat{n} = 3$, the inequality $\omega \leq \omega_c$ can be violated by several tripartite extremal no-signaling boxes (c.f. Sect. 4.4). In particular, two boxes can achieve the value $\omega = 4/3\omega_c$. It is thus clear that the bound ω_c is not a consequence of the no-signaling condition.

4.5.2 Outlook

We could check numerically that for the choice $\hat{n} = 2\lfloor \frac{n}{2} \rfloor + 1$ the inequalities $\omega \leq \omega_c$ are facets of the local polytope up to $n = 7$. Yet, they can always be violated by no-signalling correlations. This game thus tightly identifies part of the boundary separating quantum from supra-quantum correlations.

4.6 Simulating Projective Measurements on the GHZ State

One way to study nonlocal correlations is to try to simulate them with a measurable amount of nonlocal resources. This allows one to put an upper bound on the power of these correlations. For instance, it is well known that correlations created upon measurement of a singlet state can be simulated by the use of shared randomness supplemented by 1 bit of communication [39], or by 1 use of a PR box [40]. Thus no correlations found upon measurement of a singlet state can achieve a task that 1 bit of communication, or 1 PR box, cannot.

Here, we consider the simulation of the n-partite Greenberger-Horne-Zeilinger (GHZ) state

$$|GHZ\rangle = \frac{1}{\sqrt{2}}(|00\ldots0\rangle + |11\ldots1\rangle). \tag{4.6.1}$$

Since this state is genuinely tripartite nonlocal (it can violate the Svetlichny inequality (4.1.1) or its n-partite generalization [15, 16]), it cannot be simulated with just shared randomness and interaction between a subset of the parties. Nevertheless, we consider the task of simulating it with the aid of bipartite resources only.

4.6.1 Nonlocal Resources

Let us allow the parties to share nonlocal boxes of the following kinds in addition to pre-established randomness.

PR box. A Popescu-Rohrlich (PR) box [41] is a nonlocal box that admits two bits $x, y \in \{0, 1\}$ as inputs and produces locally random bits $a, b \in \{0, 1\}$, which satisfy the binary relation

$$a + b = xy. \tag{4.6.2}$$

M box. A Millionaire box [42] is a nonlocal box that admits two continuous inputs $x, y \in [0, 1]$ and produces locally random bits $a, b \in \{0, 1\}$, such that the following relation is satisfied:

$$a + b = \mathrm{sg}(x - y) \tag{4.6.3}$$

where addition is modulo 2 and the sign function is defined as $\mathrm{sg}(x) = 0$ if $x > 0$ and $\mathrm{sg}(x) = 1$ if $x \leq 0$.

Note that none of these boxes is signaling: the outcomes produced by the boxes are locally random and thus carry no information on the other party's choice of input.

4.6.2 Simulation

A protocol to simulate measurements on the GHZ state (4.6.1) with nonlocal boxes runs as follows: before letting the parties choose their measurement settings, they are allowed to share any information, plus a number of boxes (as shown in Fig. 4.4). The parties can then choose their measurement settings (which we represent by vectors $\vec{a}$, $\vec{b}$, $\vec{c}$ on the Bloch sphere). They are allowed to locally process this setting together with the pre-established shared randomness and accesses to their boxes. The parties then output the result of this process, which we denote by $\alpha, \beta, \gamma \in \{-1, 1\}$.

Using the above boxes, we considered the simulation of the correlations found by measuring the GHZ state in the equatorial plane, i.e. with $\vec{a} = (\cos\phi_a, \sin\phi_a, 0)$, etc. In this case the correlations take the form

$$\langle\alpha\rangle = \langle\beta\rangle = \ldots = \langle\alpha\beta\rangle = \ldots = 0 \tag{4.6.4}$$

for all marginal correlations, and

$$\langle\alpha\beta\ldots\rangle = \cos(\phi_a + \phi_b + \ldots) \tag{4.6.5}$$

for the full n-partite correlation term. Here are the results that we could show (proofs in [43]):

Theorem *Equatorial von Neumann measurements on the tripartite GHZ state can be simulated with 2 M boxes and 2 PR boxes distributed as in Fig. 4.4.*

Theorem *Equatorial von Neumann measurements on the 4-partite GHZ state can be simulated with 4 M boxes and 4 PR boxes distributed as in Fig. 4.5.*

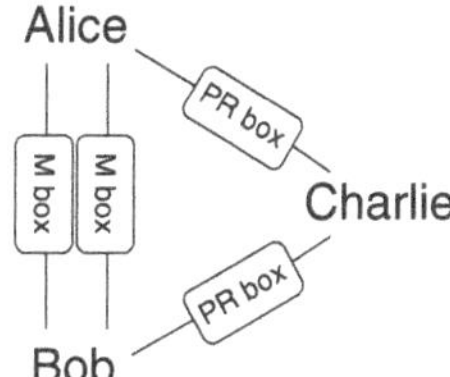

Fig. 4.4 Setup for the simulation of the tripartite GHZ state in the $x-y$ plane: two Millionaire boxes are shared between Alice and Bob and each of them shares a PR box with Charlie

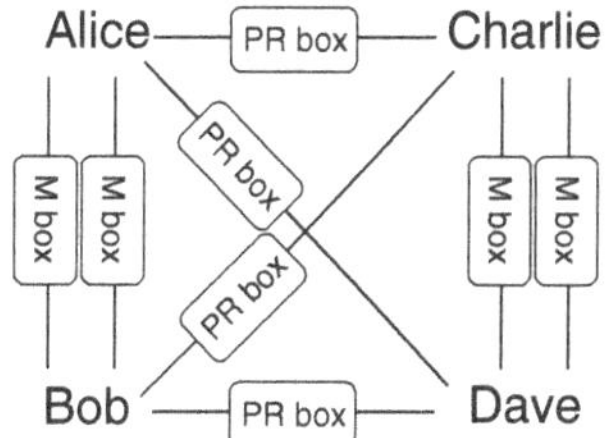

Fig. 4.5 Distribution of bipartite no-signalling boxes that allows for the simulation of equatorial von Neumann measurement on the 4-partite GHZ state

Moreover, one can show that a PR box can be simulated with one bit of classical communication transmitted from one end of the box to the other one, and an M box with an average of 4 bits. Each of these protocols can thus be translated into communication models with a finite-average communication cost. Namely, and average of 10 bits allow for the simulation of equatorial measurements on the tripartite GHZ state, whereas 20 bits suffice on average for the four-partite case.

4.6.3 Conclusion

We showed that bipartite no-signalling resources are enough to reproduce the nonlocal character of these GHZ correlations, even though these correlations are genuinely multipartite nonlocal. Moreover, we provided models to reproduce these correlations with a finite amount of communication on average. Note that this latter result was recently improved for the tripartite case [44].

References

1. O. Gühne, G. Tóth, Phys. Rep. **474**, 1 (2009)
2. G. Svetlichny, Phys. Rev. D **35**, 3066 (1987)
3. J. Barrett, S. Pironio, J.-D. Bancal, N. Gisin, The definition of multipartite nonlocality. Phys. Rev. A 88, 014102 (2013) arXiv:1112.2626
4. R. Gallego, L.E. Würflinger, A. Acín, M. Navascués, An operational framework for nonlocality Phys. Rev. Lett. 109, 070401 (2012) arXiv:1112.2647
5. I. Pitowski, *Quantum Probability-Quantum Logic* (Springer, Berlin, 1989)
6. R.F. Werner, M.M. Wolf, Phys. Rev. A **64**, 032112 (2001)
7. M. Żukowski, Č. Brukner, Phys. Rev. Lett. **88**, 210401 (2002)
8. J.-D. Bancal, N. Gisin, S. Pironio, J. Phys. A: Math. Theor. **43**, 385303 (2010)
9. J.F. Clauser, M.A. Horne, A. Shimony, R.A. Holt, Phys. Rev. Lett. **23**, 880 (1969)
10. P.M. Pearle, Phys. Rev. D **2**, 1418 (1970)
11. S.L. Braunstein, C.M. Caves, Ann. Phys. (N.Y.) **202**, 22 (1990)
12. D. Collins, N. Gisin, N. Linden, S. Massar, S. Popescu, Phys. Rev. Lett. **88**, 040404 (2002)
13. D. Kaszilkowski, L.C. Kwek, J.-L. Chen, M. Żukowski, C.H. Oh, Phys. Rev. A **65**, 032118 (2002)
14. J. Barrett, A. Kent, S. Pironio, Phys. Rev. Lett. **97**, 170409 (2006)
15. D. Collins, N. Gisin, S. Popescu, D. Roberts, V. Scarani, Phys. Rev. Lett. **88**, 170405 (2002)
16. M. Seevinck, G. Svetlichny, Phys. Rev. Lett. **89**, 060401 (2002)
17. J.-L. Chen, D.-L. Deng, H.-Y. Su, C. Wu, C.H. Oh, Phys. Rev. A **83**, 022316 (2011)
18. J.-D. Bancal, N. Brunner, N. Gisin, Y.-C. Liang, Phys. Rev. Lett. **106**, 020405 (2011)
19. J.-D. Bancal, C. Branciard, N. Brunner, N. Gisin, Y.-C. Liang, J. Phys. A: Math. Theor. **45**, 125301 (2012)
20. L. Aolita, R. Gallego, A. Cabello, A. Acín, Phys. Rev. Lett. **108**, 100401 (2012)
21. N.D. Mermin, Phys. Rev. Lett. **65**, 1838 (1990)
22. S.M. Roy, V. Singh, Phys. Rev. Lett. **67**, 2761 (1991)
23. M. Ardehali, Phys. Rev. A **46**, 5375 (1992)
24. A.V. Belinskiĭ, D.N. Klyshko, Phys. Usp. **36**, 363 (1993)
25. N. Gisin, H. Bechmann-Pasquinucci, Phys. Lett. A **246**, 1 (1998)

26. K.F. Pal, T. Vértesi, Phys. Rev. A **83**, 062123 (2011)
27. R.F. Werner, M.M. Wolf, Phys. Rev. A **61**, 062102 (2000)
28. A. Klyachko, quant-ph/0409113 (2004)
29. J. Barrett, N. Linden, S. Massar, S. Pironio, S. Popescu, D. Roberts, Phys. Rev. A **71**, 022101 (2005)
30. R. Horodecki, P. Horodecki, M. Horodecki, Phys. Lett. A **200**, 340 (1995)
31. L.E. Würflinger, J.-D. Bancal, A. Acín, N. Gisin, T. Vértesi, Phys. Rev. A **86**, 032117 (2012)
32. http://typo.zib.de/opt-long_projects/Software/Porta/
33. S. Pironio, J.-D. Bancal, V. Scarani, J. Phys. A: Math. Theor. **44**, 065303 (2011)
34. C. Śliwa, Phys. Lett. A **317**, 165 (2003)
35. T. Fritz, Polyhedral duality in Bell scenarios with two binary observables. J. Math. Phys. **53**, 072202 (2012) arXiv:1202.0141
36. B.S. Cirel'son, Lett. Math. Phys. **4**, 93 (1980)
37. G. Brassard, H. Buhrman, N. Linden, A.A. Méthot, A. Tapp, F. Unger, Phys. Rev. Lett. **96**, 250401 (2006)
38. M. Pawłowski, T. Paterek, D. Kaszlikowski, V. Scarani, A. Winter, M. Żukowski, Nature **461**, 1101–1104 (2009)
39. B.F. Toner, D. Bacon, Phys. Rev. Lett. **91**, 187904 (2003)
40. N.J. Cerf, N. Gisin, S. Massar, S. Popescu, Phys. Rev. Lett. **94**, 220403 (2005)
41. S. Popescu, D. Rohrlich, Found. Phys. **24**, 379 (1994)
42. A.C.C. Yao, in *23rd Annual Symposium on Foundations of Computer Science, Chicago*. IEEE, New York, 1982, p. 160
43. J.-D. Bancal, C. Branciard, N. Gisin, Adv. Math. Phys. **2010**, Article ID 293245 (2010)
44. C. Branciard, N. Gisin, Phys. Rev. Lett. **107**, 020401 (2011)

Chapter 5
Quantifying Multipartite Nonlocality

By performing local measurements on an n-partite entangled state one obtains outcomes that may be nonlocal, in the sense that they violate a Bell inequality [1]. Since the seminal work of Bell, nonlocality has been a central subject of study in the foundations of quantum theory and has been supported by many experiments [2, 3]. More recently, it has also been realized that it plays a key role in various quantum information applications [4, 5], where it represents a resource different from entanglement.[1]

While nonlocality has been extensively studied in the bipartite ($n = 2$) and to a lesser extent in the tripartite ($n = 3$) case, the general n-partite case remains much unexplored. The physics of many-particle systems, however, is well known to differ fundamentally from the one of a few particles and to give rise to new interesting phenomena, such as phase transitions or quantum computing. Entanglement theory, in particular, appears to have a much more complex and richer structure in the n-partite case than it has in the bipartite setting [6, 7]. This is reflected by the fact that multipartite entanglement is a very active field of research that has led to important insights into our understanding of many-particle physics (see, e.g., [8, 9]). In view of this, it seems worthy to investigate also how nonlocality manifests itself in a multipartite scenario. What new features emerge in this context and what are their fundamental implications? How to characterize the nonlocality of experimentally realizable multi-qubit states, such as W states for instance? What role do n-partite nonlocal correlations play in quantum information protocols, e.g., in measurement-based computation [10]?.

The vision behind the present paper is that in order to answer such questions and make further progress on our understanding of multipartite nonlocality, one should first find ways to *quantify* it. Motivated by this idea, we introduce two simple measures that quantify the multipartite extent of nonlocality.

---

[1] This chapter appeared as: "J.-D. Bancal, et al., *Quantifying Multipartite Nonlocality*, Phys. Rev. Lett. **103**, 090503 (2009)."

J.-D. Bancal, *On the Device-Independent Approach to Quantum Physics*, Springer Theses, DOI: 10.1007/978-3-319-01183-7_5,

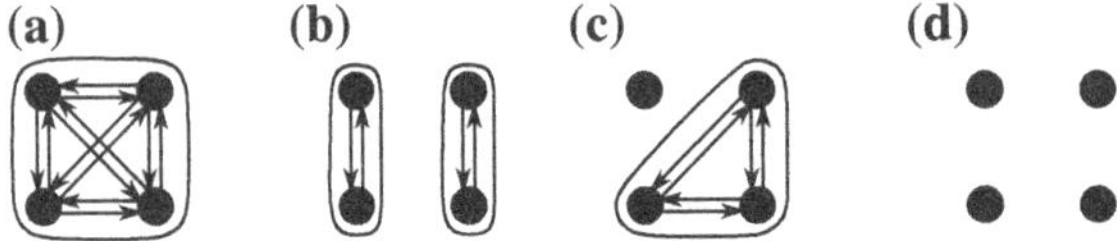

Fig. 5.1 Different groupings of $n = 4$ parties into m groups. Within each group, every party can communicate to any other party, as indicated by the *arrows*. **a** If all parties join into one group ($m = 1$), they can achieve any correlations. **b, c** If they split into $m = 2$ groups they can realize some non-local correlations but not all. **d** If they are all separated ($m = n$), they can only reproduce local correlations

A natural way to characterize nonlocality is to attempt to replicate it using models where some non-local interactions (such as communication) are allowed between some parties. The first measure that we consider is based on classical communication models à la Svetlichny [11–14], where the n parties are divided into m disjoint subgroups. Within each group the parties are free to collaborate and communicate with each other, but are not allowed to do so between distinct groups. The idea is that a given set of correlations contains more multipartite nonlocality if more parties need to join to be able to reproduce these correlations (see Fig. 5.1). The second measure of multipartite nonlocality that we introduce is based on models where k parties broadcast their measurement inputs to all others. The idea again is that correlations that require more broadcasting parties to be simulated contain more multipartite nonlocality. The maximal number m of groups and the minimal number k of broadcasting parties that allow for the reproduction of a given set of correlations thus represent two simple ways of quantifying their multipartite nonlocal content.

Given an arbitrary set of correlations, it may in general be difficult to determine the corresponding values of m and k. To evaluate these quantities, we introduce a family of Bell tests based on the Mermin-Svetlichny (MS) inequalities [11, 15]. Specifically, we compute the maximal value of the Mermin-Svetlichny (MS) expressions achieved by models where n parties form m groups and where k parties broadcast their inputs. By comparing the amount by which quantum states violate the MS inequalities with our bounds, one thus obtains constraints on the values of m and k necessary to reproduce their nonlocal correlations. Since these criteria are based on Bell-like inequalities they can be tested experimentally.

A Bell-like test for a given number of groups could a priori depend on how the groups are formed, e.g., $2 + 2$ in Fig. 5.1b or $1 + 3$ in Fig. 5.1c, and on which party belongs to which group. But, the tests that we present here depend only on the total number m of groups, and not on how the parties are distributed within each group. Furthermore, in the measurement scenario that we consider in this work (restricted to "correlation functions"), a communication model with m disjoint groups is less powerful than a communication model with $k = n - m$ broadcasting parties. Yet, we find that the bounds on the MS expressions are identical in both cases.

As mentioned above, our results can be used to estimate the multipartite nonlocal content of quantum states. We carry out this analysis for GHZ-like and W states in the last part of this paper.

Definitions We consider a Bell experiment involving n parties which can each perform one out of two measurements. The outcomes of these measurements are written a_j and a'_j and can take the values ± 1. Letting $M_1 = a_1$, we define recursively the MS polynomials [11, 12, 15, 16] as

$$M_n = \frac{1}{2}(a_n + a'_n)M_{n-1} + \frac{1}{2}(a_n - a'_n)M'_{n-1} \tag{5.1}$$

$$M_n^{\pm} = \frac{1}{\sqrt{2}}\left(M_n \pm M'_n\right), \tag{5.2}$$

where M'_n is obtained from M_n by exchanging all primed and non-primed a_j's. M_n^+ and M_n^- are equivalent under the exchange $\{a_j, a'_j\} \leftrightarrow \{-a'_j, a_j\}$ for any single party j, which corresponds to a relabeling of its inputs and outputs. The MS polynomials are symmetric under permutations of the parties.

We interpret these polynomials as sums of expectation values by identifying each term of the form $a_1 \ldots a_n$ with the correlation coefficient $\langle a_1 \ldots a_n \rangle$, which is the expectation value of the product of the outputs $a_1 \ldots a_n$. The above polynomials can thus be interpreted as Bell inequalities. Their local bounds are known [12] to be $|M_n| \leq 1$ and $|M_n^{\pm}| \leq \sqrt{2}$, while the algebraic bounds (the maximal value achieved by an arbitrary nonlocal model) are easily found to be $|M_n| \leq 2^{\lfloor \frac{n}{2} \rfloor}$ and $|M_n^{\pm}| \leq 2^{\lfloor \frac{n-1}{2} \rfloor + \frac{1}{2}}$.

In the remainder of this paper, we shall be interested in the following family of polynomials:

$$S_n^m = \begin{cases} M_n & \text{for } n - m \text{ even} \\ M_n^+ & \text{for } n - m \text{ odd}. \end{cases}$$

Quantifying multipartite nonlocality through communication models. In a classical communication model, the n parties have access to shared randomness and are allowed to communicate their inputs to some other parties. Given the information available to them, each party then produces a local output. Here, as explained in the introduction, we define two families of models that depend on a parameter m (or $k = n - m$) which quantify the extent of multipartite nonlocality.

- ***Grouping***: The n parties are grouped into m subsets. Within each group, the parties are free to collaborate with each other, but are not allowed to do so between distinct groups.
- ***Broadcasting***: Out of the n parties, k of them can broadcast their input to all other parties. The remaining $m = n - k$ parties cannot communicate their input to any other party.

In the framework of these two communication models, the values that can be reached by the MS polynomials are bounded as follows:

Theorem For the grouping and the broadcasting models,

$$|S_n^m| \leq 2^{(n-m)/2}. \quad (5.3)$$

Moreover this bound is tight, i.e., for each model there exists a strategy that yields $|S_n^m| = 2^{(n-m)/2}$ (in the case of the grouping model, this is true for any possible grouping of the n parties into m groups).

Before proving our theorem, let us elaborate on some comments. First of all, let us mention that, for $m = 2$, the results obtained in [12, 13] for the grouping model are recovered.

Note also that since we consider correlation functions only, the grouping model is weaker than the broadcasting model. Indeed, in each group one can assume that all parties send their inputs to one singled-out party, which decides for the correlation function of the whole group. The broadcasting model clearly allows more communication than this.

The fact that the same bounds hold for the two models is not trivial and is actually a special property of the MS expressions. Indeed, we have been able to construct inequalities that distinguish between these models.

Also, note that since the above bounds are tight for both models, bounds for any intermediary model, in which for instance some parties join to form groups and other broadcast their inputs, can be readily computed.

A more technical remark. As observed in [14] for the case $m = 2$, the structure of the MS inequalities allows to detect a stronger form of non-locality than the one induced by grouping. It is interesting to identify precisely the most general communication model associated with this stronger form of nonlocality.

The common feature of the two above models that we exploit in our proof (see below) and that fundamentally limits the values of the S_n^m expressions is that in both cases, there exists a special subset of m parties such that none of the n parties knows more than one input from this subset. This is obvious in the case of the broadcasting model; in the case of the grouping model, simply pick one party in each of the m groups. Let us therefore define the most general (but less natural) communication model with this property:

• ***Restrained-subset model***: Among the n parties, there is a subset of m parties, such that none of the n parties receive more than one input from this subset. The other parties are free to communicate as they wish. Note that the parties within the special subset of m parties cannot receive inputs from any other party in the subset, as they already know their own input.

This model also satisfies the bound (5.3); for the case $m = 2$, the results of [14] are recovered. This model is optimal for the MS expressions, in the sense that any additional communication between the parties allows them to violate (5.3).

Proof of (3). It is sufficient to prove (5.3) for our strongest model, i.e., for the restrained-subset model. Since the MS inequalities are symmetric under permutations of the parties, we can assume without loss of generality that the parties $1, \ldots, m$ are the ones in the restrained subset.

Consider first the case where $n-m$ is even, for which $S_n^m = M_n$. Applying twice the recursive definition (5.1), we get

$$M_n = \tfrac{1}{2}(a_n a_{n-1} M'_{n-2} + a_n a'_{n-1} M_{n-2} + a'_n a_{n-1} M_{n-2} - a'_n a'_{n-1} M'_{n-2})\,. \tag{5.4}$$

Using again twice (5.1) for $M^{(\prime)}_{n-2}$, we can replace $M^{(\prime)}_{n-2}$ as a function of $M^{(\prime)}_{n-4}$ in (5.4). Iterating this process $\frac{n-m}{2}$ times, we end up with the following expression for M_n:

$$M_n = \frac{1}{2^{(n-m)/2}} \sum_{s_n,\dots,s_{m+1}=0}^{1} a_n^{s_n} \dots a_{m+1}^{s_{m+1}} M_m^{s_n\dots s_{m+1}}, \tag{5.5}$$

where $a_i^0 = a_i$, $a_i^1 = a'_i$, and where, depending on the value of $(s_n, \dots, s_{m+1})$, $M_m^{s_n,\dots,s_{m+1}}$ is equal to one of the polynomials $\{\pm M_m, \pm M'_m\}$.

The MS polynomial $M_m^{s_n\dots s_{m+1}}$ is a function of the outputs $\{a_1, a'_1 \dots, a_m, a'_m\}$, i.e. $M_m^{s_n\dots s_{m+1}} = M_m^{s_n\dots s_{m+1}}(a_1, a'_1 \dots, a_m, a'_m)$. Among the parties $\{m+1, \dots, n\}$ there exists a (possibly empty) subset $\{j_1, \dots, j_l\}$ that do not receive any input from parties $2, \dots, m$, but possibly from party 1. Define two effective outputs A_1 and A'_1 as $A_1 = a_1 a_{j_1}^{s_{j_1}} \dots a_{j_l}^{s_{j_l}}$ and $A'_1 = a'_1 a_{j_1}^{s_{j_1}} \dots a_{j_l}^{s_{j_l}}$. There also exist similar disjoint subsets for parties $2, \dots, m$, for which we also define effective outputs $A_2, A'_2, \dots, A_m, A'_m$. Then we can write

$$a_n^{s_n} \dots a_{m+1}^{s_{m+1}} M_m^{s_n\dots s_{m+1}} = M_m^{s_n\dots s_{m+1}}(A_1, A'_1 \dots, A_m, A'_m).$$

Formally, $M_m^{s_n\dots s_{m+1}}(A_1, A'_1 \dots, A_m, A'_m)$ is a MS polynomial that involves m parties isolated from each other, since the outputs A_j, A'_j of party j do not depend on the input of any of the other $m-1$ parties. It can therefore not exceed its local bound 1. Inserting this bound in (5.5), we find $|M_n| \le 2^{(n-m)/2}$.

For odd values of $n-m$, we have to consider the polynomials $S_n^m = M_n^+$. Using the definitions (5.1) and (5.2), one can show that M_n^+ has a similar decomposition as M_n in (5.5). The same reasoning as before then leads to $|M_n^+| \le 2^{(n-m)/2}$. □

Proof of the tightness of (3). To prove that (5.3) is a tight bound, it is sufficient to prove that it can be reached by our weaker communication model, ie the grouping model (for any possible distribution of the n parties into m groups).

Let G_i $(i = 1, \dots, m)$ denote the m groups into which the n parties are split. For all group G_i having an odd number n_i of parties, there exists a strategy for the parties in G_i to reach both algebraic bounds $|M_{G_i}| = |M'_{G_i}| = 2^{(n_i-1)/2}$ at the same time. This is because the (tight) algebraic bound for $M_{G_i}^+$ is $2^{n_i/2}$ and Eq. (5.2) tells us that in order to achieve it both M_{G_i} and M'_{G_i} must reach their algebraic limit. Similarly there exists a strategy for groups with even number of parties n_i such that

$|M^+_{G_i}| = |M^-_{G_i}| = 2^{(n_i-1)/2}$. We shall thus associate $\{M, M'\}$ polynomials to odd groups and $\{M^+, M^-\}$ to even ones.

Consider two groups G_i, G_j, and their union $G_{ij} = G_i \cup G_j$. From the definitions (5.1) and (5.2), one can derive the decompositions:

$$
\begin{aligned}
M_{G_{ij}} &= \frac{1}{2}\left[M_{G_i}\left(M_{G_j} + M'_{G_j}\right) + M'_{G_i}\left(M_{G_j} - M'_{G_j}\right)\right] \\
M_{G_{ij}} &= \frac{1}{2}\left[M^+_{G_i}\left(M^+_{G_j} + M^-_{G_j}\right) + M^-_{G_i}\left(M^+_{G_j} - M^-_{G_j}\right)\right] \\
M^\pm_{G_{ij}} &= \frac{1}{2}\left[M^\pm_{G_i}\left(M_{G_j} + M'_{G_j}\right) \mp M^\mp_{G_i}\left(M_{G_j} - M'_{G_j}\right)\right].
\end{aligned}
$$

Similar relations are also obtained for $M'_{G_{ij}}$ since $(M^\pm_G)' = \pm M^\pm_G$. Now inserting in the above relations the value attained by the strategies that we just mentioned for the two initial groups, one finds that their combined strategy can achieve $|M_{G_{ij}}| = |M'_{G_{ij}}| = 2^{(n_i-1)/2}\, 2^{(n_j-1)/2}$ or $|M^+_{G_{ij}}| = |M^-_{G_{ij}}| = 2^{(n_i-1)/2}\, 2^{(n_j-1)/2}$, depending on which set of polynomials are associated to the two initial groups.

Iterating this construction by joining groups successively 2 by 2, we find $|M_n| = \prod_{i=1}^m 2^{(n_i-1)/2}$ when there is an even number of even groups, and $|M^+_n| = \prod_{i=1}^m 2^{(n_i-1)/2}$ otherwise. Since the parity of the number of even groups is the same as the parity of $n-m$, there must exists a strategy which achieves $|S_n| = \prod_{i=1}^m 2^{(n_i-1)/2}$ $= 2^{(n-m)/2}$. □

Nonlocality of quantum states. Suppose that one observes a violation of the inequality $|S^m_n| \leq 2^{(n-m)/2}$. One can then conclude that in order to reproduce the corresponding nonlocal correlations in the framework of our communication models, the parties cannot be separated in more than $m-1$ groups, or that at least $k+1 = n-m+1$ parties must broadcast their input. Thus, the above bounds on S^m_n give us bounds on the multipartite character of the observed nonlocal correlations (an upper bound on m, or a lower bound on k).

Here we discuss the violation of the inequalities (5.3) for n-partite GHZ-like and W states. States in the GHZ family are defined as $GHZ_\theta\rangle = \cos\theta|00\ldots0\rangle + \sin\theta|11\ldots1\rangle$. The maximal value of M_n for these states was conjectured in [17] to be $M_n = \max\{1, 2^{(n-1)/2}\sin 2\theta\}$. Numerical optimizations (see Fig. 5.2) induce us to conjecture that similarly $M^+_n = \max\{\sqrt{2}, 2^{(n-1)/2}\sin 2\theta\}$. Upon comparison with the bound (5.3), we conclude that all n-partite GHZ states with $\theta > \pi/8$ are maximally non-local according to our criterion (i.e., all parties must be grouped together or $n-1$ parties must broadcast their input to reproduce their correlations). Less entangled GHZ states, on the other hand, cannot be simulated if the parties are separated in more than $m-1$ groups or if fewer than $k+1 = n-m+1$ parties broadcast their inputs whenever $\theta > \theta_c$ with $\sin 2\theta_c = 2^{-(m-1)/2}$. Interestingly, θ_c is the same for all n.

Consider now the W states $|W_n\rangle = \frac{1}{\sqrt{n}}(|10\ldots0\rangle + \cdots + |0\ldots01\rangle)$. Numerical optimizations suggest that the maximal values of the MS polynomials for these states are upper-bounded by a small constant for all n (see Fig. 5.3). To convince ourselves

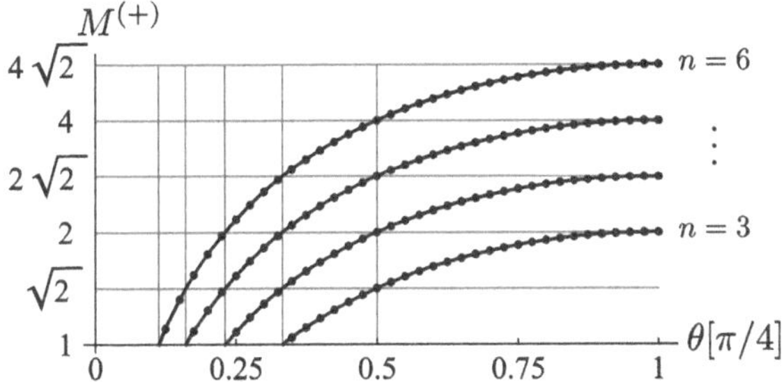

Fig. 5.2 Maximal values of M_n and M_n^+ for partially entangled GHZ states for $3 \leq n \leq 6$. The *dots* are values found by numerical optimization and the *solid lines* are the conjectured violation $M_n = M_n^+ = 2^{(n-1)/2} \sin 2\theta$ (valid only above 1 for M_n and $\sqrt{2}$ for M_n^+)

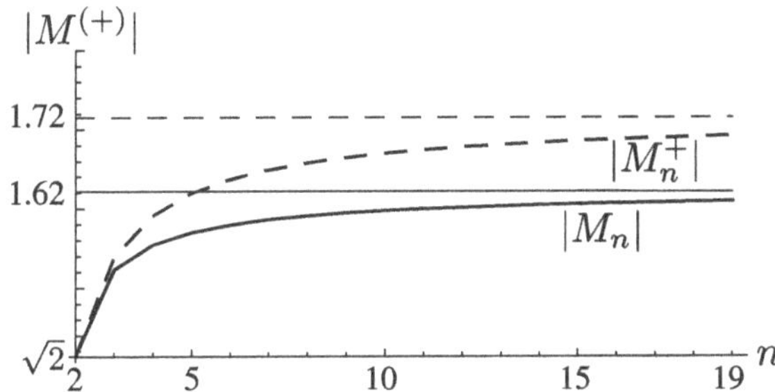

Fig. 5.3 Maximal values of M_n (*solid line*) and M_n^+ (*dashed line*) for n-partite W states. The *curves* were obtained by a general numerical optimization for $n \leq 9$, and under the hypothesis that all parties use identical measurement settings for $10 \leq n \leq 19$. The asymptotic values for $n \to \infty$ computed as explained in the text are also shown

that this is indeed the case, we analyzed analytically the case where all pairs of measurement settings are the same for all parties. This is justified by the results of our numerical optimizations up to $n = 9$, for which the optimal measurement settings can always be of this form. We thus introduce for all n parties two measurement operators A_0 and A_1 represented by vectors $a_i = (\sin\theta_i \cos\phi_i, \sin\theta_i \sin\phi_i, \cos\theta_i)$. One can show that as n increases, the maximal value of $|M_n|$ or $|M_n^+|$ can be reached for $\phi_i = 0$ and $\theta_i \to 0$. Assuming a power law for $\theta_i(n)$, one finds that it should be given by $\theta_i \sim c_i/\sqrt{n}$ at the maximum. After optimization of the constants c_0 and c_1 for both M_n and M_n^+, we found that the asymptotic maximal values of the MS polynomials (under our assumptions, which we believe are not restrictive) are $|M_\infty| \simeq 1.62$, $|M_\infty^+| = 2\sqrt{2/e}$. Since $S_n^{n-1} > 1$ for $n \geq 3$, letting one party broadcast his input, or letting two parties join to form a group, is not sufficient to reproduce the correlations of the W state. However, we cannot reach the same conclusion if more than two parties join or if $k = n - m \geq 2$ parties broadcast their inputs, since the criterion (5.3) is not violated in this case.

Conclusion We have proposed in this paper two simple measures of multipartite nonlocality and have introduced a series of Bell tests to evaluate them. This represents a primary step towards a quantitative understanding of quantum nonlocality for an arbitrary number n of parties.

While GHZ states exhibit a strong form of multipartite nonlocality according to our criterion, we have found that W states violate our inequalities only for small values of k. This suggest that W states exhibit only a very weak form of multipartite non-locality. Or, it might actually be that other inequalities are necessary to quantify properly the nonlocality of W states. Finding which one of these possibilities is the correct one is an interesting problem for future research. Also, it would be interesting to analyze the non-locality of other kinds of multipartite quantum states with our criterions.

As suggested by the situation in entanglement theory, we do not expect our measures to be the only ways to quantify the multipartite content of nonlocality. It would thus be of interest to look for other ways to quantify multipartite nonlocality, based on other nonlocal models than the ones considered here.

Finally, let us stress that the criteria that we presented in this paper can be tested experimentally. It would thus be worth (re-)considering experiments on multipartite nonlocality in view of our results.

We acknowledge support by the Swiss NCCR *Quantum Photonics* and the European ERC-AG *QORE*.

References

1. J. Bell, *Speakable and Unspeakable in Quantum Mechanics* (Cambridge University Press, Cambridge, 1987)
2. M. Genovese, Phys. Rep. **413**, 319 (2005)
3. A. Aspect, Nature **398**, 189 (1999)
4. J. Barrett, N. Linden, S. Massar, S. Pironio, S. Popescu, D. Roberts, Phys. Rev. A **71**, 022101 (2005)
5. Č. Brukner, M. Żukowski, J.-W. Pan, A. Zeilinger, Phys. Rev. Lett. **92**, 127901 (2004)
6. W. Dür, G. Vidal, J. I. Cirac, Phys. Rev. A **62**, 062314 (2000)
7. R. Horodecki, P. Horodecki, M. Horodecki, K. Horodecki, Rev. Mod. Phys. **81**, 865 (2009)
8. A. Osterloh, L. Amico, G. Falci, R. Fazio, Nature **416**, 608 (2002)
9. G. Vidal, Phys. Rev. Lett. **93**, 040502 (2004)
10. R. Raussendorf, H.J. Briegel, Phys. Rev. Lett. **86**, 5188 (2001)
11. G. Svetlichny, Phys. Rev. D **35**, 3066 (1987)
12. D. Collins, N. Gisin, S. Popescu, D. Roberts, V. Scarani, Phys. Rev. Lett. **88**, 170405 (2002)
13. M. Seevinck, G. Svetlichny, Phys. Rev. Lett. **89**, 060401 (2002)
14. N.S. Jones, N. Linden, S. Massar, Phys. Rev. A **71**, 042329 (2005)
15. N.D. Mermin, Phys. Rev. Lett. **65**, 1838 (1990)
16. R.F. Werner, M.M. Wolf, Phys. Rev. A **64**, 032112 (2001)
17. V. Scarani, N. Gisin, J. Phys. A: Math. Gen. **34**, 6043 (2001)

Chapter 6
Device-Independent Entanglement Detection

Entanglement is one of the most intriguing feature of quantum physics. It allows several particles to be in a state which cannot be understood as a concatenation of the sate of each particle. Experimental demonstration of entanglement is generally performed with one of the two following techniques: tomography of the full quantum state, or evaluation of an entanglement witness.

In the first case, the state ρ of the system is characterized by performing a number of complementary measurements on it [1]. For instance, on two qubits, measurement of the product of all Pauli operators $\sigma_i \otimes \sigma_j$, with $j = 0, 1, 2, 3$, and $\sigma_0 = \mathbb{1}$ allows one in principle to deduce ρ by solving the set of linear equations

$$\mathrm{tr}(\rho\, \sigma_i \otimes \sigma_j) = f_{ij} \tag{6.1}$$

where f_{ij} is the observed frequency for the corresponding measurements. In practice however, experimental imperfections typically lead to a solution for this set of equations which is unphysical so that more complicated techniques are generally used instead of the linear inversion, like maximum likelihood estimation [2]. Still, once the reconstructed state is found, theoretical analyses can be performed on it to check directly whether the quantum state is entangled or not.

In contrast, an entanglement witness is an observable $\mathcal{W}$ such that $\mathrm{tr}(\rho\mathcal{W}) \geq 0$ whenever ρ is separable [3]. Any decomposition of $\mathcal{W}$ in terms of local observables allows one to evaluate it by performing local measurements on the state under consideration. If a value $\mathrm{tr}(\rho\mathcal{W}) < 0$ is found, the measured state is then said to be entangled.

6.1 Imperfect Measurements

Any experimental manipulation is affected by imperfections, be it only the finiteness of the number of times measurements are repeated in order to accumulate sufficient statistics. Interestingly, the effect that statistical uncertainties on the frequencies f_{ij}

J.-D. Bancal, *On the Device-Independent Approach to Quantum Physics*, Springer Theses, DOI: 10.1007/978-3-319-01183-7_6, © Springer International Publishing Switzerland 2014

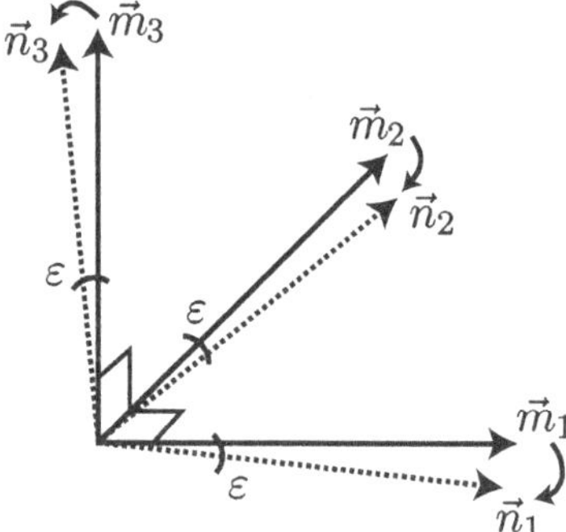

Fig. 6.1 Intended and actual measurement directions for tomography on a qubit. The actual measurement directions $\vec{n}$ are distant from the desired ones $\vec{m}$ by an angle smaller than ε

can have on tomographically reconstructed states was analysed rigorously only very recently [4, 5].

Still, even in absence of statistical uncertainties, which can in principle be avoided by performing enough measurements in a random order, systematic errors in the measurement process can also affect the conclusion of a test. While this problem is known, it is seldom discussed in the literature. Let us show what kind of effects these errors can have on the detection of entanglement.

6.1.1 Effects of Systematic Errors on Tomography

In order to evaluate the effect of systematic errors on the process of tomography reconstruction, we consider the situation in which each measurement can be slightly misaligned. Namely, if $\vec{m} \cdot \vec{\sigma}$ denotes the desired measurement on a qubit, the actual measurement performed can be written as $\vec{n} \cdot \vec{\sigma}$ with $\vec{m} \cdot \vec{n} \geq \cos(\varepsilon)$ (c.f. Fig. 6.1). However, since $\vec{n}$ is unknown, results measured along $\vec{n}$ are interpreted during the reconstruction process as coming from measurements along $\vec{m}$.

Considering qubit states $|\psi\rangle$, we looked for the maximum effect that these errors could have on the reconstructed state ρ by performing the following optimization :

$$\begin{aligned} \min_{|\psi\rangle, n_i} \quad & \langle\psi|\rho|\psi\rangle \\ \text{subject to} \quad & \vec{m}_i \cdot \vec{n}_i \geq \cos(\varepsilon) \ \forall \text{ measurement } i \end{aligned} \tag{6.2}$$

The result of this numerical optimization is shown in Fig. 6.2a. For small errors ε in the definition of the measurement bases, the uncertainty of the reconstructed n-qubit state increases at least as $\frac{n}{\sqrt{2}}\varepsilon$ (c.f. [6] for more details). Thus, if measurements are done with linear polarizers having a precision of 1° in real space, for instance, the precision of the reconstructed state can decrease by 2.5 % per qubit in the system.

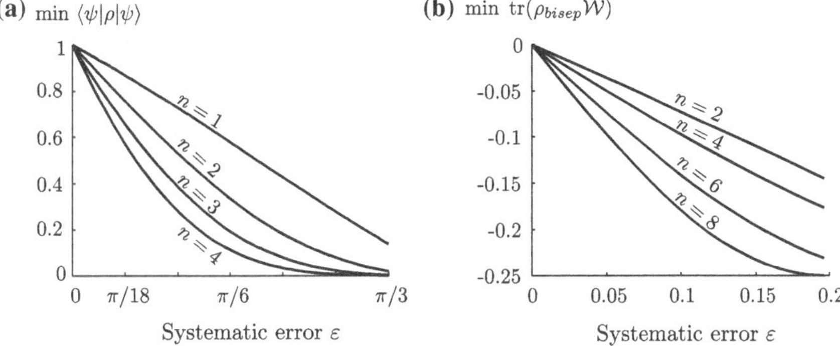

Fig. 6.2 **a** Minimum fidelity of the tomographically reconstructed n-qubit state when the measurement settings deviate by at most ε from the requested ones. **b** Minimum expectation value of $\mathrm{tr}(\mathcal{W}\rho)$ found using imperfect measurements on bispearable states

Interestingly, we found that entangled states are usually more robust to systematic errors than the worst bound shown in Fig. 6.2a (see [6]). Nevertheless, imperfect measurements on separable states can sometimes lead to an entangled reconstructed state (Rosset, private communication). Entanglement can thus be wrongly witnessed through tomography because of systematic errors.

6.1.2 *Effects of Systematic Errors on Entanglement Witnesses*

In a similar fashion, we analysed the effect of systematic errors on the witness

$$\mathcal{W} = \frac{1}{2}\mathbb{1} - |GHZ\rangle\langle GHZ| \tag{6.3}$$

which detects genuine multipartite entanglement [3]. For this we used the decomposition of $\mathcal{W}$ in terms of local operators given in [7]. Allowing again all measurement operators to differ from the prescribed ones by at most ε, we looked for the smallest value $\mathrm{tr}(\rho\mathcal{W})$ that could be achieved by measuring a biseparable state ρ.

The results are shown in Fig. 6.2b. In particular, we note that a negative value can be found, and thus entanglement wrongly detected, as soon as $\varepsilon > 0$. This entanglement witness measured in this manner is thus sensitive to misalignment of the measurements.

6.2 Witnesses Insensitive to Systematic Errors?

If the entanglement detection schemes we just mentioned are sensitive to systematic errors, these kind of errors, unlike statistical errors, can be hard to evaluate in practice: how can one make sure that measurement settings are perfectly aligned? or that they are not more misaligned than some ε? How can one certify that the measurements do not act on a larger Hilbert space than expected? etc. Even in the case that these errors can be reasonably estimated, taking them into account makes the analysis of the situation complicated... It is thus worth asking whether entanglement detection can be made resistant to systematic errors.

The answer to this question is already known since several years: Bell inequalities can detect entanglement without relying on any hypothesis about the kind of measurements performed or even about the nature of the measured system. Indeed, since no Bell inequality can be violated by measuring a separable state, violation of a Bell inequality witnesses that the state measured is entangled. All it needs to certify entanglement through the violation of a Bell inequality, in a bipartite scenario for instance, is a proper way of measuring the correlations $P(ab|xy)$. That is, the inputs of each party should be indexed by x and y, their outputs by a and b, and the systems should be well identified from each other. The conclusion is then device-independent as discussed in the introduction of this thesis.

Conversely, if the correlations found during an experiment violate no Bell inequality, then the presence of entanglement cannot be certified based only on the device-independent hypotheses. Indeed, in this case a local strategy can reproduce the correlations, and thus some measurements on a separable state as well. Entanglement can thus be demonstrated in a device-independent manner if and only if measurement statistics can violate a Bell inequality.

More precisely, if a Bell inequality $\sum_{abxy} \beta_{abxy} P(ab|xy) \leq c$ is satistifed by all local correlations, then the measurement operators $M_{a|x}$, $M_{b|y}$ used during the experiment (which might not be the ones that we think we're measuring) define the operator $\mathcal{W} = c\mathbb{1} - \sum_{abxy} \beta_{abxy} M_{a|x} \otimes M_{b|y}$, which is a witness for entanglement: it satisfies $\mathrm{tr}(\rho\mathcal{W}) \geq 0$ for every state ρ that is separable. A negative value $\mathrm{tr}(\rho\mathcal{W}) < 0$ is found as soon as the Bell inequality is violated. When testing a Bell inequality we thus test an entanglement witness which is adapted to the measurement implemented experimentally, even if these measurements are not known to us. The conclusion is thus independent of these measurements.

6.2.1 Device-Independent Witnesses for Genuine Tripartite Entanglement

The equivalence between Bell inequalities and device-independent entanglement witnesses remains true in scenarios involving more than two parties. However, in these scenarios one is typically interested in witnessing genuine multipartite

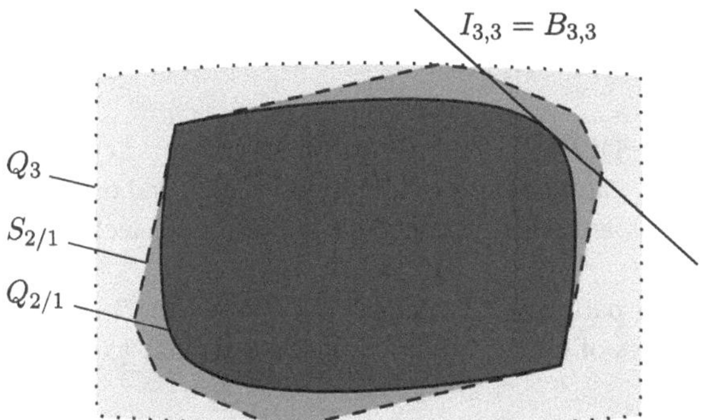

Fig. 6.3 Particular slice in the space of tripartite correlations with three settings and two outcomes representing schematically the sets of general quantum correlations (Q_3), Svetlichny correlations ($S_{2/1}$) and biseparable quantum correlations ($Q_{2/1}$). The inequality (6.5) is also represented and detects correlations that are not genuine tripartite-nonlocal

entanglement [3]. Indeed, it is generally easier to test entanglement between a subset of parties directly on those parties specifically.

If tripartite Bell inequalities can typically be violated with states that are not genuinely multipartite entanglement, and are thus unable to witness this kind of entanglement, violation of a Svetlichny inequality demonstrates genuine tripartite nonlocality (c.f. Sect. 4.1) and is thus sufficient to demonstrate genuine tripartite entanglement as well. But it is not necessary [8]. Rather, genuinely tripartite entanglement can be demonstrated from observation of correlations $P(abc|xyz)$ as soon as these correlations cannot be obtained by measuring a biseparable quantum state ρ_{bisep}, i.e. if the correlations are not *biseparable correlations*:

$$P_{bisep}(abc|xyz) = \text{tr}(\rho_{bisep} M_{a|x} \otimes M_{b|y} \otimes M_{c|z}) \tag{6.4}$$

where $M_{a|x}$, $M_{b|y}$ and $M_{c|z}$ are arbitrary measurement operators.

The set of tripartite biseparable correlations can be described in terms of a hierarchy of semidefinite programming (SDP) (see Fig. 6.3 and Chap. 7). It is thus possible to determine directly from the observed correlations whether a genuinely tripartite entangled state was measured, or whether these correlations are compatible with measurements performed on an biseparable state, in which case genuine multipartite entanglement cannot be demonstrated with the device-independent hypotheses only.

6.2.2 A Witness for Genuine Multipartite Entanglement

Letting $\vec{s} \in \{0, \ldots, m-1\}^n$ be a vector denoting the choice of measurement for the n parties within m possible ones, and $\vec{r} \in \{0, 1\}^n$ be the vector of their outcomes, the following inequality is satisfied by all biseparable quantum correlations (c.f. Sect. 4.2.2.3):

$$I_{n,m} = \sum_{[\mathbf{s}]_m=0} (-1)^{\lfloor \frac{\mathbf{s}}{m} \rfloor} E_{\vec{s}} + \sum_{[\mathbf{s}]_m=1} (-1)^{\lfloor \frac{\mathbf{s}-1}{m} \rfloor} E_{\vec{s}} \leq 2m^{n-2} \cot\left(\frac{\pi}{2m}\right) = B_{n,m} \quad (6.5)$$

where $E_{\vec{s}} = \sum_{\vec{r}} (-1)^{\mathbf{r}} P(\vec{r}|\vec{s})$ is the n-partite correlator, $[x]_m = x - \lfloor \frac{x}{m} \rfloor m$ and $\mathbf{s} = \sum_i s_i$, $\mathbf{r} = \sum_i r_i$ are the sums of all parties' inputs and outputs. This inequality can thus be used as a device-independent witness to detect genuine multipartite entanglement.

Measuring the n-partite GHZ state $|GHZ\rangle = \frac{1}{2}(|0\rangle^{\otimes n} + |1\rangle^{\otimes n})$ with the jth measurement settings of every party lying in the x-y plane as

$$\cos(\phi_j)\sigma_x + \sin(\phi_j)\sigma_y, \quad \text{with } \phi_j = -\frac{\pi}{2mn} + j\frac{\pi}{m}, \quad \text{for } j = 0, \ldots, m-1, \quad (6.6)$$

yields the value $I_{n,m} = 2m^{n-1} \cos\left(\frac{\pi}{2m}\right) > B_{n,m}$. This inequality can thus detect genuine multipartite entanglement in noisy GHZ states $\rho = V|GHZ\rangle\langle GHZ| + (1 - V)\mathbb{1}/2^n$ with visibilities down to $V_c = \left(m \sin\left(\frac{\pi}{2m}\right)\right)^{-1}$. As the number of settings m increases, this critical visibility decreases, tending to the value of $2/\pi$ (c.f. also [10]).

While inequality (6.5) reduces to the Svetlichny inequality for $m = 2$, and thus also detects genuine n-partite nonlocality in this case, it does not do so anymore for more inputs. Multipartite entanglement is then detected with a lower visibility than multipartite nonlocality.

6.3 Experimental Demonstration

In the group of Prof. R. Blatt in Innsbruck, we tested the inequality (6.5) on a system of trapped $^{40}Ca^{+}$ ions [11].

6.3.1 Experimental Setup and Procedure

We used a linear trap loaded with $n = 3, 4$ or 6 ions, the logical states $|0\rangle$ and $|1\rangle$ of each ions being encoded in the $S_{1/2}(m = -1/2)$ ground state and $D_{5/2}(m = -1/2)$ metastable state, respectively.

After initialization of the system in the ground state of the center-of-mass motion by Doppler and sideband cooling and in the logical state $|0\rangle^{\otimes n}$ by optical pumping, the ions can be brought to the GHZ entangled state by applying a Mølmer-Sørensen gate [12].

Measurement of the ions in the computational basis is achieved by the electron-shelving technique by scattering light on the $S_{1/2} \leftrightarrow P_{1/2}$ transition, and detecting the fluorescence with a photomultiplier tube. In order to perform measurements of all ions in the x-y plane of the Bloch sphere, we first apply local phase gates $\exp(-i\frac{\phi}{2}\sigma_z)$ by means of AC-stark-shift beams focused on individual ions. The x axis of the Bloch sphere of all ions is then brought to the computational basis by applying a collective $\pi/2$ rotation on the qubit transition.

While the Mølmer–Sørensen gate can yield maximally entangled states with a high fidelity, the coherence time of the GHZ state produced in this way is of about 2 ms for $n = 3$ ions, and it decreases quadratically with the number of entangled qubits [13]. The duration of the σ_z rotations, performed sequentially on the ions, taking $\sim$100 µs/2π, a significant decrease in the quality of the state can take place during the application of these pulses. To avoid this effect, we inverted the state of half of the ions before doing the Mølmen-Sørensen entangling gate for $n = 4, 6$. This allows one to produce a decoherence-free GHZ state of the form $\frac{1}{2}(|0\rangle^{\otimes n/2}|1\rangle^{\otimes n/2} + e^{i\varphi}|1\rangle^{\otimes n/2}|0\rangle^{\otimes n/2})$ whose coherence time of $\sim$300 ms leaves enough time to manipulate all ions. Note that the measurements settings (6.6) need to be adapted to this new state. We thus used the following ones on $n = 4, 6$ ions: for $m = 2$ settings, we used the phases $\phi_j = j\frac{\pi}{2}$ for the first half of the ions and $\phi'_j = \frac{n+1}{12n}\pi + \frac{1-j}{2}\pi$ for other ones; for $m = 3$ we used $\phi_j = \frac{n+1}{12n}\pi + j\frac{\pi}{3}$ for the first ions and $\phi'_j = \frac{2-j}{3}\pi$ for the other ones, where $j = 0, \ldots, m-1$ denotes the different measurement setting of each party. The optimal state then has the phase $\varphi = \frac{5-7n}{24\pi}$.

Finally, to cancel the effect of eventual drifts during the experiment, measurements sets were taken by blocks of 50 identical measurement chosen in a random order.

6.3.2 Addressing Errors

In order to show an indisputable violation of (6.5), the experiment producing the correlations should close all loopholes that can appear in a Bell experiment (c.f. Chap. 1). Of course, this is not the case in the present experiment: even though the detection loophole is closed here, measurements were not performed in a space-like manner. In fact, the different systems are not even totally isolated from each other since they are separated by only 3$\sim$5 µm. The measurements performed on the ions might thus not be put in a tensor form as assumed in Sect. 6.2.

If arbitrary joint measurements are allowed in the decomposition (6.4) instead of a tensor product of local measurement, any correlation can be obtained by measuring a biseparable state, and thus any value of the inequality (6.5) can be reached as well. No interesting bound on an inequality can thus be proven in the presence of arbitrary cross-talks, i.e. without making some assumptions about which cross-talk is present in the system. Note that this situation is similar to the fact that no interesting bound can usually be put on a standard entanglement witness in presence of arbitrary systematic errors. We thus performed a special analysis in order to estimate the amount of cross-talk in our system, and how it could influence the biseparable bound $B_{n,m}$ of the inequality.

In our system, we expect the strongest source of cross-talk between the ions to be due to the imperfect focusing of the AC-stark shift lasers. Indeed, it is the only action which is supposed to act on some ions specifically and which might not: leakage of this laser onto neighbouring ions can cause them to feel part of the rotation imposed on the first ion. The state of an ion, or equivalently the basis in which it is measured, can thus depend on the measurement settings of the other ions. This effect can be modeled by replacing the measurement phases ϕ_j by $\phi'_j = \sum_k M_{jk}\phi_k$ where $M_{jj} = 1$, and $0 \leq M_{jk} \leq \epsilon$ if $j \neq k$ is the amount of cross-talk from ion k to ion j. Here ϵ is a bound on the worst addressing error.

In order to evaluate the impact that these errors can have on the bound $B_{n,m}$, we estimated the amount of addressing errors present in the experiment. This allowed us to determine an upper bound ϵ on the addressing errors which was not exceeded in the experiment, except possibly with a probability smaller than 10^{-6}. This upper bound is $\epsilon = 1.6\,\%$ for $n = 3$, $\epsilon = 5\,\%$ for $n = 4$, and $\epsilon = 4\,\%$ for $n = 6$. We then computed numerically the maximum impact $\Delta I^{AE} = I^{\epsilon}_{bisep} - I^{\epsilon=0}_{bisep}$ that addressing errors bounded by ϵ can have on the biseparable bound for the settings we intended to use in the experiment. Assuming that the maximum contribution of the addressing errors to (6.5) is given by ΔI^{AE}, we update the bound B to $B^{AE} = B + \Delta I^{AE}$ to account for the cross-talks present in the experiment.

Note that even though the actual measurement settings might differ from the ones we intended to measure, the modified bound B^{AE} remains valid in the presence of cross-talk if the measurements implemented in the lab differ (not too much) from the ideal ones, because $B > I^{\epsilon=0}_{bisep}$.

6.3.3 Experimental Results

The experimental evaluation of the witness are summarized in Table 6.1. In all cases the measured values are consistent with a visibility of the state of about 90 %, except for the tripartite case in which the GHZ state was not decoherence-free.

Table 6.1 Summary of the experimental measurement of (6.5)

n	m	B^{AE}	I^{exp}	Visibility (%)	$I^{exp} - B^{AE}$ (σ units)
3	2	4.070(4)	4.78(6)	84(1)	12
	3	10.542(9)	12.39(1)	79.48(9)	136
4	2	8.43(8)	10.42(6)	92(1)	20
	3	32.5(2)	42.53(8)	90.9(7)	41
6	2	34.0(5)	40(1)	88(3)	5
	3	294(3)	374(3)	89(1)	19

For each scenario considered, the value of (6.5) observed is reported as I^{exp}, together with the associated visibility, i.e. the ratio between this value and the one expected from optimal measurement on a perfect GHZ state. The experimental value should be compared to the bound B^{AE}, which includes a correction due to the addressing errors observed between the ions (c.f. Sect. 6.3.2)

The inequalities with two inputs ($m = 2$) coincides with the Svetlichny inequalities and thus detect genuine multipartite nonlocality. The witness with three inputs ($m = 3$) however, is able to detect genuine multipartite entanglement even in absence of genuine nonlocality. This allows one to demonstrate stronger violations as shown in Table 6.1.

6.4 Conclusion

Any measurement of a device-independent entanglement witness results in the test of a standard entanglement witness which relies on the measurement settings actually implemented in the lab rather than on measurement settings which might not exactly correspond to the experimental situation. This ensures that a violation of the inequality cannot be caused by a miscalibration of the experiment. Device-independent witnesses are thus particularly robust to (possibly unknown) measurement imperfections inherent to every experimental test.

Motivated by this perspective, we constructed device-independent witnesses able to detect genuine multipartite entanglement. Since these witnesses can detect genuine multipartite entanglement even in absence of genuine multipartite nonlocality, they can provide larger experimental violations than tests of Svetlichny inequalities, as was demonstrated in the experiment we conducted with the Innsbruck ion group of Prof. R. Blatt.

Despite being robust to imperfect measurements, the bounds of these witnesses can be affected by cross-talks between subsystems if these are not perfectly isolated from each other. Since this problem is quite generic, and is present in many experimental setups, it deserves further investigation.

References

1. G.M. D'Ariano, M. De Laurentis, M.G.A. Paris, A. Porzio, S. Solimeno, J. Opt. B: Quantum Semiclass. Opt. **4**, S127 (2002)
2. Z. Hradil, J.ehéek, J. Fiuráek, M. Jeek, Lecture Notes, in *Physics: Quantum State Estimation*, ed. by M.G.A. Paris, J. ehéek, (Springer, Berlin, 2004), pp. 59–112
3. O. Gühne, G. Tóth, Physics Reports **474**, 1 (2009)
4. Matthias Christandl and Renato Renner, Reliable Quantum State Tomography Phys. Rev. Lett. **109**, 120403 (2012) arXiv:1108.5329.
5. Robin Blume-Kohout, Robust error bars for quantum tomography (Submitted on 23 Feb 2012) arXiv:1202.5270.
6. D. Rosset, R. Ferretti-Schöbitz, J.-D. Bancal, N. Gisin, Y.-C. Liang, Phys. Rev. A **86**, 062325 (2012)
7. O. Gühne, C.-Y. Lu, W.-B. Gao, J.-W. Pan, Phys. Rev. A 76, 030305(R) (2007)
8. J.L. Cereceda, Phys. Rev. A **66**, 024102 (2002)
9. J.-D. Bancal, C. Branciard, N. Brunner, N. Gisin, Y.-C. Liang, J. Phys. A: Math. Theor **45**, 125301 (2012)

10. K.F. Pal, T. Vértesi, Phys. Rev. A **83**, 062123 (2011)
11. F. Schmidt-Kaler, H. Häffner, S. Gulde, R. Riebe, G.P.T. Lancaster, T. Deuschle, C. Becher, W. Hänsel, J. Eschner, C.F. Roos, R. Blatt, Appl. Phys. B **77**, 789–796 (2003)
12. G. Kirchmair, J. Benhelm, F. Zähringer, R. Gerritsma, C.F. Roos, R. Blatt, New J. Phys. **11**, 023002 (2009)
13. T. Monz, P. Schindler, J.T. Barreiro, M. Chwalla, D. Nigg, W.A. Coish, M. Harlander, W. Hänsel, M, Hennrich and R. Blatt. Phys. Rev. Lett. **106**, 130506 (2011)

Chapter 7
Device-Independent Witnesses of Genuine Multipartite Entanglement

We consider the problem of determining whether genuine multipartite entanglement was produced in an experiment, without relying on a characterization of the systems observed or of the measurements performed. We present an n-partite inequality that is satisfied by all correlations produced by measurements on biseparable quantum states, but which can be violated by n-partite entangled states, such as Greenberger-Horne-Zeilinger states. In contrast to traditional entanglement witnesses, the violation of this inequality implies that the state is not biseparable independently of the Hilbert space dimension and of the measured operators. Violation of this inequality does not imply, however, genuine multipartite nonlocality. We show more generically how the problem of identifying genuine tripartite entanglement in a deviceindependent way can be addressed through semidefinite programming.

The generation of multipartite entanglement is a central objective in experimental quantum physics. For instance, entangled states of fourteen ions and six photons have recently been produced [1, 2]. In any such experiment, a typical question arises: How can we be sure that genuine n-partite entanglement was present? A state is said to be genuinely n-partite entangled if it is not biseparable, that is, if it cannot be prepared by mixing states that are separable with respect to some partition. Consider for instance the tripartite case: a state ρ_{bs} is said to be biseparable if it admits a decomposition

$$\rho_{\text{bs}} = \sum_k \rho_{AB}^k \otimes \rho_C^k + \sum_k \rho_{AC}^k \otimes \rho_B^k + \sum_k \rho_{BC}^k \otimes \rho_A^k, \tag{7.1}$$

where the weight of each individual state in the mixture has been included in its normalization; a state that cannot be written as above is genuinely tripartite entangled. Determining whether genuine n-partite entanglement was produced in an experiment represents a difficult problem that has attracted much attention recently (see e.g. [3–6]). The usual approach consists of measuring a witness of genuine multipartite

J.-D. Bancal, *On the Device-Independent Approach to Quantum Physics*, Springer Theses, DOI: 10.1007/978-3-319-01183-7_7, © Springer International Publishing Switzerland 2014

entanglement, or of doing the full tomography of the state followed by a direct analysis of the reconstructed density matrix.[1]

Such approaches, however, not only rely on the observed statistics to conclude about the presence of entanglement, but also require a detailed characterization of the systems observed and of the measurements performed. Consider for instance the following witness of genuine tripartite entanglement:

$$M = X_1X_2X_3 - X_1Y_2Y_3 - Y_1X_2Y_3 - Y_1Y_2X_3, \tag{7.2}$$

where $X_j = \sigma_x$ and $Y_j = \sigma_y$ are the Pauli spin observables in the x and y direction for particle j. For any biseparable three-qubit state $\langle M\rangle = \text{tr}\,(M\rho) \leq 2$ [7]. Thus if we measure three spin-$\frac{1}{2}$ particles in the x and y direction and find an average value $\langle M\rangle > 2$, we can conclude that the state exhibits genuine tripartite entanglement.

Suppose, however, that the measurement Y_3 carries a slight (possibly unnoticed) bias towards the x direction. That is, instead of measuring $Y_3 = \sigma_y$, we actually measure $Y_3 = \cos\theta\sigma_y + \sin\theta\sigma_x$. Then it is not difficult to see, all other measurements being ideal, that the biseparable state $|\psi\rangle = \frac{1}{2}\left(|00\rangle + e^{-i\phi}|11\rangle\right)_{AB} \otimes (|0\rangle + |1\rangle)_C$, where $\phi = \arctan(\sin\theta)$, yields $\langle M\rangle = 2\sqrt{1+\sin^2\theta}$ which is strictly larger than 2 for any $\theta \neq 0$. Thus, unless we measure all particles *exactly* along the x and y directions, we can no longer conclude that observing $\langle M\rangle > 2$ implies genuine tripartite entanglement. Importantly, this is not a unique feature of the above witness, but rather all conventional witnesses are, to some extent, susceptible to such systematic errors that are seldom taken into account.

Furthermore, tomography and usual entanglement witnesses typically assume that the dimension of the Hilbert space is known. For instance, in a typical experiment demonstrating, say, entanglement between four ions, we usually view each ion as a two-level system. But an ion is a complex object with many degrees of freedom (position, vibrational modes, internal energy levels, etc). How do we know, given the inevitable imperfections of the experiment, that it is justified to treat the relevant Hilbert space of each ion as two-dimensional and how does this simplification affects our conclusions about the entanglement present in the system [8]? Even if it is justified to view each ion as a qubit, is entanglement between four systems really necessary to reproduce the measurement data, or could they be reproduced with fewer entangled systems if qutrits were manipulated instead?

These remarks motivate the introduction of entanglement witnesses that are able to guarantee that a quantum system exhibits (n-partite) entanglement, without relying on the types of measurements performed, the precision involved in their implementation, or on assumptions about the relevant Hilbert space dimension. We call such witnesses, *device-independent* entanglement witnesses (DIEW). This type of approach was already considered in Refs. [9–12]. Note that other solutions to the aboveproblems are possible, such as entanglement witnesses tolerating

[1] This chapter appeared as "J.-D. Bancal, et al., *Device-Independent Witnesses of Genuine Multipartite Entanglement*, Phys. Rev. Lett. **106**, 250404 (2011)."

a certain misalignement in the measurement apparatuses [13] or the characterization of realistic measurement apparatuses through squashing maps [8]. These types of more specific approaches, however, still require some partial characterization of the system and measurement apparatuses, which is not necessary when using DIEWs.

Any DIEW is a Bell inequality (i.e. a witness of nonlocality). Indeed, (i) the violation of a Bell inequality implies the presence of entanglement, and (ii) any measurement data that does not violate any Bell inequality can be reproduced using quantum states that are fully separable [14]. The violation of a Bell inequality is thus a necessary and sufficient condition for the detection of entanglement in a device-independent (DI) setting. This observation is the main insight behind DI quantum cryptography [15, 16], where the presence of entanglement is the basis of security.

The relation between DIEW for *genuine n-partite* entanglement and witnesses of multipartite nonlocality is more subtle. While there exist Bell inequalities that detect genuine n-partite nonlocality [17–19], not every DIEW for n-partite entanglement corresponds to such a Bell inequality. Consider for instance, the expression (7.2). If no assumptions are made on the type of systems observed and measurements performed, the inequality $\langle M \rangle \leq 2$ corresponds to Mermin's Bell-type inequality [20], i.e., a value $\langle M \rangle > 2$ necessarily reveal non-locality, hence entanglement. Moreover, a value $\langle M \rangle > 2\sqrt{2}$ guarantees genuine tripartite entanglement [9, 18]. The Mermin expression (7.2) can thus be used as a tripartite DIEW. Yet, it cannot be used as a Bell inequality for genuine tripartite non-locality, since a simple model involving communication between two parties only already achieves the algebraic maximum $\langle M \rangle = 4$ [18].

The objectives of this paper are to formalize the concept of DIEW for genuine multipartite entanglement and initiate a systematic study that goes beyond the early examples given in [9–12]. Following this line, we start by introducing the notion of quantum biseparable correlations. We then present a simple DIEW for n-partite entanglement which is stronger for GHZ (Greenberger-Horne-Zeilinger) states than all the inequalities introduced in [9–12]. In the case $n = 3$, we also provide a general method for determining whether given correlations reveal genuine tripartite entanglement and apply it to GHZ and W states. Apart from yielding practical criteria for the characterization of entanglement in a multipartite setting, our results also clarify the relation between device-independent multipartite entanglement and multipartite nonlocality.

Biseparable quantum correlations.—For simplicity of exposition, let us consider an arbitrary tripartite system (the following discussion easily generalizes to the n-party case). To characterize in a DI way its entanglement properties, we consider a Bell-type experiment: on each subsystem, one of m possible measurements is performed, yielding one of d possible outcomes. We adopt a black-box description of the experiment and represent the measurements on each of the three subsystems by classical labels $x, y, z \in \{1, \ldots, m\}$ (corresponding, e.g., to the values of macroscopic knobs on the measurement apparatuses) and denote the corresponding classical outcomes $a, b, c \in \{1, \ldots, d\}$. The correlations obtained in the experiment are characterized by the joint probabilities $P(abc|xyz)$ of finding the triple of outcomes a, b, c given the measurement settings x, y, z.

We say that the correlations $P(abc|xyz)$ are *biseparable quantum correlations* if they can be reproduced through local measurements on a biseparable state ρ_{bs}. That is, if there exist a biseparable quantum state (7.1) in some Hilbert space $\mathcal{H}$, measurement operators $M_{a|x}$, $M_{b|y}$, and $M_{c|z}$ (which without loss of generality we can take to be projections satisfying $M_{a|x}M_{a'|x} = \delta_{a,a'}M_{a|x}$ and $\sum_a M_{a|x} = \mathbb{1}$), such that

$$P(abc|xyz) = \text{tr}\left[M_{a|x} \otimes M_{b|y} \otimes M_{c|z}\,\rho_{\text{bs}}\right]. \tag{7.3}$$

If given quantum correlations $P(abc|xyz)$ are not biseparable, they necessarily arise from measurements on a genuinely tripartite entangled state, and this conclusion is independent of any assumptions on the type of measurements performed or on the Hilbert space dimension.

Equivalently, biseparable quantum correlations can be defined as those that can be written in the form

$$\begin{aligned} P(abc|xyz) = &\sum_k P_Q^k(ab|xy)P_Q^k(c|z) \\ &+ \sum_k P_Q^k(ac|xz)P_Q^k(b|y) \\ &+ \sum_k P_Q^k(bc|yz)P_Q^k(a|x), \end{aligned} \tag{7.4}$$

where $P_Q^k(ab|xy)$ and $P_Q^k(c|z)$ correspond, respectively, to arbitrary two-party and one-party quantum correlations, i.e., they are of the form $P_Q^k(ab|xy) = \text{tr}[M_{a|x}^k \otimes M_{b|y}^k\,\rho_{AB}^k]$ and $P_Q^k(c|z) = \text{tr}[M_{c|z}^k\,\rho_C^k]$ for some unormalized quantum states ρ_{AB}^k, ρ_C^k and measurement operators $M_{a|x}^k$, $M_{b|y}^k$, $M_{c|z}^k$ [and similarly for the other terms in (7.4)]. Here the measurement operators for different bi-partitions do not need to be the same (though this can always be achieved as shown in Sect. D of [21]). Clearly, from the definition 7.1) of biseparable states, any correlations of the form (7.3) are of the form (7.4). Conversely, any correlations of the form (7.4) are also of the form (7.3), see Sect. A of [21].

Let Q_3 denote the set of tripartite quantum correlations and $Q_{2/1} \subset Q_3$ the set of biseparable quantum correlations. From (7.4), it is clear that $Q_{2/1}$ is convex and that its extremal points are of the form $P_Q^{ext}(ab|xy)P_Q^{ext}(c|z)$ where $P_Q^{ext}(ab|xy)$ is an extremal point of the set Q_2 of bipartite quantum correlations and $P_Q^{ext}(c|z)$ an extremal point of the set Q_1 of single-party correlations (the extreme points of Q_1 are actually classical, deterministic points). Since the set $Q_{2/1}$ is convex, it can be entirely characterized by linear inequalities. Those linear inequalities separating $Q_{2/1}$ from Q_3 correspond to DIEWs for genuine tripartite entanglement. Since $Q_{2/1}$ has an infinite number of extremal points, there exist an infinite number of such inequalities. Note that the set of local correlations $P(abc|xyz) = \sum_k P^k(a|x)P^k(b|y)P^k(c|z)$ is contained in $Q_{2/1}$. This implies that any DIEW for genuine tripartite entanglement is a Bell inequality (though not necessarily a tight Bell inequality). Note also that the decomposition (7.4) corresponds to a Svetlichny-type decomposition [17] where all bipartite factors are restricted to be quantum, whereas less restrictive constraints (or even none in Svetlichny original definition $P(abc|xyz) = \sum_k P^k(ab|xy)P^k(c|z) + \ldots$) are imposed on these bipartite terms in the definitions of multipartite non-locality [17]. It follows that the set of genuinely bipartite non-local correlations is larger than the set of biseparable quantum correlations as illustrated in Fig. 7.1. Thus while any Bell inequality detecting genuine tripartite non-locality is a DIEW for genuine tripartite entanglement, the converse is not necessarily true. All these observations extend to the n-party case.

A DIEW for n-partite entanglement.—We now present a DIEW for n parties, where each party i performs a measurement $x_i \in \{1, 2, 3\}$ and obtains an outcome $a_i \in \{-1, 1\}$. We denote $E(\bar{x})$ the correlator associated to the measurement settings $\bar{x} = (x_1, \ldots, x_n)$, i.e. the expectation value $E(\bar{x}) = \sum_{\bar{a}} P(\bar{a}|\bar{x}) \prod_{i=1}^{n} a_i$, where $\bar{a} = (a_1, \ldots, a_n)$ denotes an n-tuple of outcomes. Let $E_n^k = \sum_{\bar{x}} \delta(\sum_{i=1}^{n} x_i = k)E(\bar{x})$ be the sum of correlators $E(\bar{x})$ for which the measurement settings x_i of the n parties sum up to k. Let f_k be a function such that $f_{k+3} = -f_k$ and taking successively the values $[1, 1, 0]$ on the integers $k = 0, 1, 2$. Then the inequality

$$I_n = \sum_{k=n}^{3n} f_{k-n}\, E_n^k \leq 2 \times 3^{n-3/2} \tag{7.5}$$

is satisfied by all biseparable quantum correlations, and is thus a DIEW for genuine n-partite entanglement. The proof of this statement is based on the decomposition (7.4) and is given in Sect. B of [21] The Svetlichny bound associated to the expression

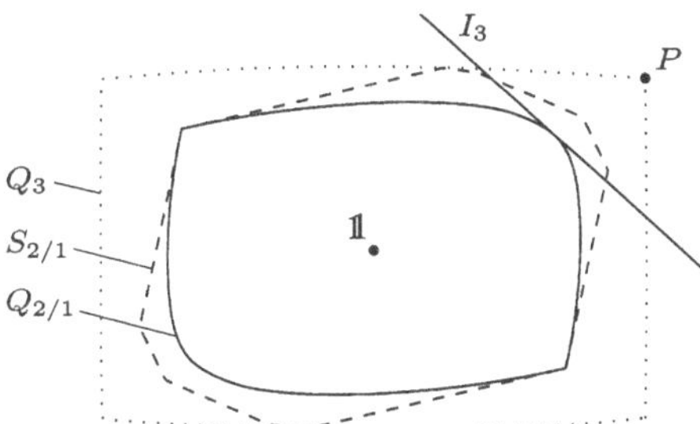

Fig. 7.1 A particular slice of the space of tripartite correlations with 3 settings and 2 outcomes representing schematically the sets of general quantum correlations ($Q3$), Svetlichny correlations ($S_{2/1}$) and biseparable quantum correlations ($Q_{2/1}$). The point $\mathbb{1}$ corresponds to random correlations and P to the GHZ correlations maximally violating the DIEW (7.5), which is represented by the straight line I_3; note that a DIEW can be violated by Svetlichny-local correlations. (The Svetlichny polytope $S_{2/1}$ can be determined exactly using linear programming, while Q_3 and $Q_{2/1}$ can be approximated efficiently using SDP techniques, see main text)

I_n, on the other hand, is easily found to be $4\times 3^{n-2} > 2\times 3^{n-3/2}$ (see Sect. B of [21]); for the local bound of I_n, see [22].

We now illustrate how this DIEW can be used to detect genuine multipartite entanglement. For this, let us consider the noisy GHZ state $\rho = V|GHZ\rangle_{n\,n}\langle GHZ| + (1-V)\mathbb{1}/2^n$ characterized by the visibility V. Carrying out the measurements $\cos((\frac{x_i-1}{3}-\frac{1}{6n})\pi)\,\sigma_x+\sin((\frac{x_i-1}{3}-\frac{1}{6n})\pi)\,\sigma_y$ on all parties, we obtain $I_n = 3^{n-1/2}V$, which violates (7.5) provided that $V > 2/3$. The DIEW (7.5) can thus detect in a DI way genuine n-partite entanglement in a noisy GHZ state for visibilities as low as $V = 2/3$. This significantly improves over the threshold visibility $V = 1/\sqrt{2}$ required to violate the DIEW based on the Mermin expression (7.2) or the different inequalities introduced in [10–12]. Note that in a DI setting it is not possible to detect genuine multipartite entanglement in tripartite GHZ states below $V = 1/2$ using projective measurements. Indeed, we have shown in this case that there exists a biseparable model reproducing all GHZ correlations (see Sect. C of [21]).

In the case $n = 3$, the DIEW (7.5) takes the form $I_3 = E_3^3+E_3^4-E_3^6-E_3^7+E_3^9 \le 6\sqrt{3}$. It therefore involves only 18 expectation values, compared to 27 for a full tomography of a three-qubit system. Let us stress, however, that contrary to usual entanglement witnesses I_3 is not restricted to two-dimensional Hilbert spaces, even though it uses observables with binary outcomes. For instance, if all parties perform the measurements $2|\phi(x_i)\rangle\langle\phi(x_i)|-\mathbb{1}$ with $|\phi(x_i)\rangle = \frac{1}{\sqrt{2}}\left(|0\rangle + e^{i(6x_i-7)\pi/18}|1\rangle\right)$ on the three-qutrit state $\frac{1}{\sqrt{3}}(|000\rangle+|111\rangle+|222\rangle)$, then $I_3 = 6\sqrt{3}+8/3$, showing that the state is genuinely tripartite-entangled.

General characterization of biseparable quantum correlations in the case $n = 3$. Though the DIEW (7.5) seems particularly well adapted to GHZ states, we cannot expect a single nor a finite set of DIEW to completely characterize the biseparable region, as illustrated in Fig. 7.1. It is thus desirable to derive a general method to decide whether arbitrary correlations are biseparable. Here we show how the semidefinite programming (SDP) techniques introduced in [23–25] can be used to certify that the correlations observed in an experiment are genuinely tripartite entangled.

Our approach is based on the observation that the tensor product separation $\rho_{AB}\otimes\rho_C$ at the level of states cf. in the definition (7.1) of biseparable states can be replaced by a commutation relation at the level of operators. Specifically, let $s = \{AB/C, AC/B, BC/A\}$ denotes the three possible partitions of the parties into two groups. Then, $P(abc|xyz)$ are biseparable quantum correlations if and only if there exist three arbitrary (not necessarily biseparable) states ρ^s and three sets of measurement operators $\{M^s_{a|x}, M^s_{b|y}, M^s_{c|z}\}$ such that

$$P(abc|xyz) = \sum_s \mathrm{tr}[M^s_{a|x}\otimes M^s_{b|y}\otimes M^s_{c|z}\rho^s], \tag{7.6}$$

where measurement operators corresponding to an isolated party commute, i.e., $[M^{AB/C}_{c|z}, M^{AB/C}_{c'|z'}] = 0$, and similarly for the other partitions. The equivalence between (7.3) and (7.6) is established in Sect. D. of [21]. The problem of determining whether given correlations $P(abc|xyz)$ are biseparable thus amounts to finding a set

Table 7.1 Summary of numerical investigations

State	$V_{\min}$ with two settings	$V_{\min}$ with three settings
GHZ	$0.7071 \simeq 1/\sqrt{2}$	$0.6667 \simeq 2/3$
W	$0.7500 \simeq 3/4$	0.7158

of operators satisfying a finite number of algebraic relations (the projection defining relations of the type $M^s_{a|x} M^s_{a'|x} = \delta_{a,a'} M^s_{a|x}$, $\sum_a M^s_{a|x} = \mathbb{1}$ and the commutation relations mentioned above) such that (7.6) holds. Such a problem is a typical instance of the SDP approach introduced in [23–25] (see details in Sect. D of [21]). Specifically, it follows from the results of [25] that one can define an infinite hierarchy of criteria that are necessarily satisfied by any correlations of the form (7.6) and which can be tested using SDP. If given correlations do not satisfy one of these criteria, we can conclude that they reveal genuinely tripartite entanglement. Further, it is possible in this case to derive an associated DIEW from the solution of the dual SDP. Modulo a technical assumption, it can be shown that the hierarchy of SDP criteria is complete, that is, if given correlations are not biseparable this will necessarily show-up at some finite step in the hierarchy.

Application to GHZ and W states.—Using finite levels of this hierarchy and optimizing over the possible measurements, we investigated the minimal visibilities above which the GHZ state $|GHZ\rangle$ and the W state $|W\rangle = (|001\rangle + |010\rangle + |100\rangle)/\sqrt{3}$ exhibit correlations that are not biseparable (and thus reveal genuine tripartite entanglement) in the case of two and three measurement settings per party. Our results are summarized in Table 7.1. For GHZ states, the reported visibility $V_{\min} = 2/3$ for three measurements per party correspond to the threshold required to violate the DIEW (7.5), suggesting that this DIEW is optimal in this case. In the case of two measurements per party, we could not lower the visibility below the threshold $V_{\min} = 1/\sqrt{2}$, which corresponds to the visibility required to violate the DIEW based on Mermin expression (7.2) and the DIEWs introduced in [9–12]. Note, however, that for $V > 1/\sqrt{2}$ the GHZ state violates Svetlichny's inequality [17] and thus exhibits genuine tripartite nonlocality. Thus for GHZ state the DIEWs introduced in [9–12] do not improve over what can already be concluded using the standard notion of tripartite non-locality. On the other hand, our numerical explorations suggest that the visibilities $V_{\min} = 2/3$ for GHZ states with three measurements and $V_{\min} = 3/4$ for W states with two measurements cannot be attained using the notion of genuine tripartite non-locality, illustrating the interest of the weaker notion of DIEW.

Dicussion.—To conclude, we comment on some possible directions for future research. First of all, note that by identifying the measurement settings "X_i" with $x_i = 1$ and "Y_i" with $x_i = 2$, the two-setting DIEW based on Mermin expression (7.2) can be written as $E^3_3 - E^5_3 \leq 2\sqrt{2}$, which is of the same general form as the three-setting DIEW (7.5). This suggests that the DIEWs based on (7.2) and (7.5) actually form part of a larger family of m-settings DIEWs. This question deserves further investigation. A second problem is to derive simple DIEWs that are adapted to W states and that can in particular reproduce the threshold visibilities obtained in Table 7.1. Finally, we have shown a practical method to characterize three-partite biseparable correlations using SDP. It would be interesting to understand how to generalize these results to the n-partite case. A possibility would be to combine the approach of [23–25] with the symmetric extensions introduced in [26, 27]. This question will be investigated elsewhere.

Acknowledgments This work was supported by the Swiss NCCRs QP and QSIT, the European ERC-AG QORE, and the Brussels-Capital region through a BB2B grant.

References

1. T. Monz et al., Phys. Rev. Lett. **106**, 130506 (2011)
2. W. Wieczorek et al., Phys. Rev. Lett. **103**, 020504 (2009)
3. O. Gühne, G. Tóth, Phys. Rep. **474**, 1 (2009)
4. O. Gühne, M. Seevinck, New J. Phys. **12**, 053002 (2010)
5. B. Jungnitsch, T. Moroder, O. Gühne, Phys. Rev. Lett. **106**, 190502 (2011)
6. M. Huber et al., Phys. Rev. Lett. **83**, 022329 (2011)
7. G. Tóth, O. Gühne, M. Seevinck, J. Uffink, Phys. Rev. A **72**, 014101 (2005)
8. T. Moroder et al., Phys. Rev. A **81**, 052342 (2010)
9. M. Seevinck, J. Uffink, Phys. Rev. A **65**, 012107 (2001)
10. K. Nagata, M. Koashi, N. Imoto, Phys. Rev. Lett. **89**, 260401 (2002)
11. J. Uffink, Phys. Rev. Lett. **88**, 230406 (2002)
12. M. Seevinck, G. Svetlichny, Phys. Rev. Lett. **89**, 060401 (2002)
13. M. Seevinck, J. Uffink, Phys. Rev. A **76**, 042105 (2007)
14. R.F. Werner, Phys. Rev. A **40**, 4277 (1989)
15. A. Acín et al., Phys. Rev. Lett. **98**, 230501 (2007)
16. S. Pironio et al., Nature (London) **464**, 1021 (2010)
17. G. Svetlichny, Phys. Rev. D **35**, 3066 (1987)
18. D. Collins et al., Phys. Rev. Lett. **88**, 170405 (2002)
19. J.-D. Bancal, N. Brunner, N. Gisin, Y.-C. Liang, Phys. Rev. Lett. **106**, 020406 (2011)
20. N.D. Mermin, Phys. Rev. Lett. **65**, 1838 (1990)
21. See supplemental material at http://link.apa.org/supplemental/10.1103/PhysRevLett. 106250404 for our biseparable model and the proofs related to Eqs. (4)–(6)
22. M. Żukowski, D. Kaszlikowski, Phys. Rev. A **56**, R1682 (1997)
23. M. Navascués, S. Pironio, A. Acín, Phys. Rev. Lett. **98**, 010401 (2007)
24. M. Navascués, S. Pironio, A. Acín, New J. Phys. **10**, 073013 (2008)
25. S. Pironio, M. Navascués, A. Acín, SIAM J. Optim. **20**(5), 2157 (2010)
26. B.M. Terhal, A.C. Doherty, D. Schwab, Phys. Rev. Lett. **90**, 157903 (2003)
27. A.C. Doherty, P.A. Parrilo, F.M. Spedalieri, Phys. Rev. Lett. **88**, 187904 (2002)

Chapter 8
Quantum Information Put into Practice

Allowing information to be carried by physical systems described by the rules of quantum physics led to a deep questioning of the theory of information. While many questions remain open, the emerging field of quantum information already led to several remarkably concrete applications which would not exist otherwise.

Here we present a modest contribution to the analysis of the security of quantum key distribution (QKD), as well as a protocol which can be used to question a database with some level of security.

8.1 Memoryless Attack on the 6-State QKD Protocol

Quantum key distribution (QKD) allows two parties who share an initial secret key of finite size, to increase its size by exchanging quantum and classical signals through an untrusted environment. The new key generated in this way can then be used for any cryptographic application [1], such as secure transmission of a secret message, a task which is not known to be possible by classical means.

Standard security proofs for QKD protocols aim at relying on the weakest possible assumptions. For instance, it is usually admitted that a possible eavesdropper is not constrained by technological limitations but only by the laws of physics. Such assumptions allow one to derive strong security bounds. However, if a particular circumstance happens to restrict further the possible action of an eavesdropper, more refined security analyses taking these limitations into account can allow the trusted parties to improve the efficiency of their protocol.

Motivated by the effort put in several groups worldwide [2–5] to implement quantum memories preserving coherence and population over more than several miliseconds, we consider the case in which the eavesdropper has no access to a long-lasting quantum memory.

Security proofs applicable in this scenario have been presented in [6] for the BB84 protocol, and more recently for the BB84, SARG and 6-state QKD protocols

J.-D. Bancal, *On the Device-Independent Approach to Quantum Physics*, Springer Theses, DOI: 10.1007/978-3-319-01183-7_8, © Springer International Publishing Switzerland 2014

[7]. Here we give a tighter bound than [7] for the achievable secure key-rate of the prepare-and-measure 6-state protocol when the eavesdropper has no access to any quantum memory.

8.1.1 The 6-State Protocol

The 6-state protocol for quantum key distribution [8] runs in 4 parts.

Distribution: Alice prepares one of the six qubit states $\rho_i = |\psi_i\rangle\langle\psi_i|$ chosen uniformly at random within

$$|\psi_1\rangle = |0\rangle,\ |\psi_2\rangle = |1\rangle,\ |\psi_3\rangle = \frac{|0\rangle + |1\rangle}{\sqrt{2}},\ |\psi_4\rangle = \frac{|0\rangle - |1\rangle}{\sqrt{2}},$$
$$|\psi_5\rangle = \frac{|0\rangle + i|1\rangle}{\sqrt{2}},\ |\psi_6\rangle = \frac{|0\rangle - i|1\rangle}{\sqrt{2}}. \tag{8.1.1}$$

She remembers the basis $b^A = \lfloor \frac{i-1}{2} \rfloor$ corresponding to this state as well as the bit $X = i - 1 \mod 2$. Alice sends this state to Bob through a public quantum channel. Upon receival of the system from Alice, Bob measures it in either the x, y, or z basis. He remembers his choice of basis $b^B = 0, 1, 2$ as well as the result of his measurement $Y = 0, 1$. This step is repeated N times, allowing the parties to accumulate the strings $\{b_k^A\}$, $\{X_k\}$ and $\{b_k^B\}$, $\{Y_k\}$.

Sifting: Alice and Bob publicly announce their choice of bases b_k^A and b_k^B. Having learned the other party's choice of basis, they discard the runs k in which $b_k^A \neq b_k^B$ (X_k and Y_k are not expected to be correlated in this case), and keep the results from the other runs indexed by k'. The basis information is then $b_{k'} = b_{k'}^A = b_{k'}^B$.

Error correction: An error correction protocol is run from Alice to Bob[1] in order to correct for expected errors between their sifted raw key strings $\{X_{k'}\}$ and $\{Y_{k'}\}$. This corrects Bob's string $\{Y_{k'}\}$ to let him hold the same sifted bit string $\{X_{k'}\}$ as Alice. This procedure also lets Alice and Bob evaluate the average Quantum Bit Error Rate (QBER): $Q = P(X_{k'} \neq Y_{k'})$.

Secure key extraction: Privacy amplification is performed on the corrected bit strings $\{X_{k'}\}$ in order to extract its secret part.

During the whole protocol, exchanges of classical information are authenticated with the initial secret key shared by the two parties in order to avoid man-in-the-middle attacks.

[1] Note that reverse reconciliation, in which Bob sends information to Alice for her to recover Bob's key, or two-way reconciliation [9] is also possible, but we don't consider this case here.

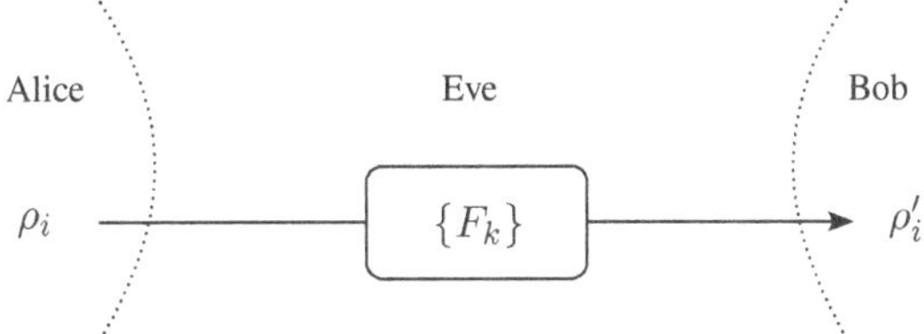

Fig. 8.1 Schematic representation of a prepare and measure QKD protocol: Alice prepares a quantum state ρ_i that she sends to Bob though a public quantum channel, which can be under the control of an eavesdropper

8.1.2 Secret Key Rate

Here, we consider an eavesdroper, Eve, which can access the quantum channel used by Alice to send the quantum states she prepares to Bob, and which can listen to all classical transmissions taking place during the protocol. However, Eve cannot hold any quantum information. Her most general interaction with the quantum channel can thus be modeled by a POVM acting independently on each of the qubits sent by Alice (c.f. Fig. 8.1).

Notice that Eve's power is greatly reduced here compared to the case in which she performs a general coherent attack. In particular she cannot use any information about the basis used by Alice or Bob $s_k^{A,B}$ to choose how to measure her system. Moreover, since each run k of the protocol is treated independently of the precedent ones by Alice and Bob, the most powerful attack that Eve can perform is an individual attack.

We thus use the Csiszár-Körner formula [10, 11], which expresses the secret key rate that Alice and Bob can extract during a realization of the protocol:

$$r = I(A:B) - \min(I(A:E), I(B:E)) \tag{8.1.2}$$

where A represents Alice's sifted key (i.e. $\{X_{k'}\}$), B Bob's sifted key, and E any system hold by Eve. $I(X:Y)$ here stands for the mutual information between variables X and Y. The mutual information between Alice and Bob is given as usual by:

$$I(A:B) = 1 - h(Q), \tag{8.1.3}$$

where $h(p) = -p\log p - (1-p)\log(1-p)$ is the binary entropy function of p. The following result provides a lower bound on the key rate r by upper-bounding the maximal mutual information between Alice and Eve as a function of the QBER.

Result *The maximum information that an eavesdropper without quantum memory can have in common with Alice's bits after sifting is given by:*

$$I(A{:}E) = \frac{1}{3}\left[1 - h\left(\frac{1-\sqrt{3Q(2-3Q)}}{2}\right)\right]. \tag{8.1.4}$$

(proof in Appendix B)

Note that this bound does not refer to the length N of the raw key produced by Alice and Bob. It is thus only valid in the limit of infinite key length $N \to \infty$.

8.1.3 Discussion

A result similar to the one presented here was recently published by Bocquet et al. in [7]. However, their analysis refers to the entanglement-based realization of the 6-state protocol. In this version, preparation of the state ρ_i by Alice is realized by letting her measure in the σ_x, σ_y or σ_z basis a maximally entangled state shared with Bob. Since the eavesdropper can interact with the quantum channel during the distribution of the entangled state ρ_{AB}, she can in principle hold a purification $|\psi_{ABE}\rangle$ such that $\mathrm{tr}_E(|\psi_{ABE}\rangle\langle\psi_{ABE}|) = \rho_{AB}$ of this state.

A comparison between the achievable key rate in the above prepare-and-measure and in the entanglement-based scheme is shown in Fig. 8.2. This shows that the key rate is slightly higher in the prepare-and-measure scheme. This contrasts with the same bounds for the BB84 QKD protocol, which are identical for both implementations.

8.2 Private Database Queries

While QKD allows one to secure the communication between two trustfull parties, many more cryptographic tasks can be considered. Here we consider a situation in which Alice wants to learn about an element of a database held by Bob, without letting Bob know which element she's interested in.

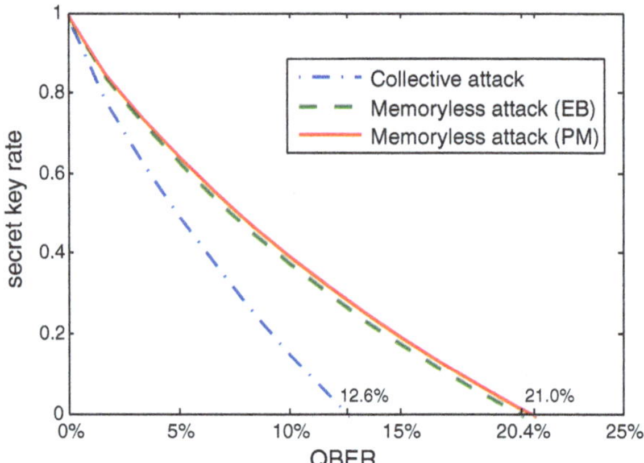

Fig. 8.2 Comparison of the secret key rate of the 6-state protocol in different situations. The bound for collective attack is as given by [12]. The two bounds against adversary without a quantum memory are in the entanglement-based scheme (EB) as given by [7] and as given by Eq. (8.1.4) for the prepare and measure scheme (PM)

This task is also known as 1 out of N oblivious transfer (for a database of N elements) [13]. The security of the database querries consists of two parts:

- Database security: Bob wants to bound the information that Alice can access on his database. Ideally he would like this information to be restricted to 1 bit per querry.
- User privacy: Alice wants to bound the probability that Bob can learn which item of his database she is interested in. Ideally, he should get no information about it.

Even though it was proved that both aspects of the security cannot be fully satisfied at the same time [14, 15], Giovannetti et al. [16] recently proposed a quantum protocol that could provide a reasonable level of security for both the user and the database provider.

However, in lossy situations the security of their protocol is compromised: since it requires Alice to send her question to Bob before knowing whether her system will come back with an answer or not, Bob can take advantage of the losses by requiring Alice to send her question several times, and thus learning what her question is with high probability.

Here we propose a protocol for private database queries based on the SARG QKD protocol [17], which is fundamentally noise-resistant. After presenting the main ideas of our protocol, we argue about the partial security it provides to both the database provider, and the user.

8.2.1 Sketch of the Protocol

The protocol for private database queries presented here is based on the SARG04 QKD protocol [17], and only differs in the classical processing. We summarize here the main steps of the protocol.

Distribution: Bob uniformly chooses one of the four qubit-states $|\uparrow\rangle$, $|\rightarrow\rangle$, $|\downarrow\rangle$, $|\leftarrow\rangle$ and sends it to Alice. Alice measures the quantum system she received from Bob either in the σ_x or in the σ_z basis and records the measured state.

Sifting: If Alice didn't receive some systems from Bob, due to losses, she tells so to Bob which discards these runs. This allows the protocol to be loss resistant since at this stage, no information about the database, or about Alice's question has been exchanged. For the systems that Alice received, Bob announces a pair of states within the following ones which contains the state he prepared: $\{|\uparrow\rangle, |\rightarrow\rangle\}$, $\{|\rightarrow\rangle, |\downarrow\rangle\}$, $\{|\downarrow\rangle, |\leftarrow\rangle\}$, $\{|\leftarrow\rangle, |\uparrow\rangle\}$.

Transcoding: Bob translates the state $|\uparrow\rangle$ and $|\downarrow\rangle$ to bits 0, and $|\leftarrow\rangle$ and $|\rightarrow\rangle$ to bits 1. On her side, knowing her measurement results as well as the sifting sets, Alice tries to guess the bit that Bob computed. This can be summarized in the following table if her measurement result is $|\uparrow\rangle$:

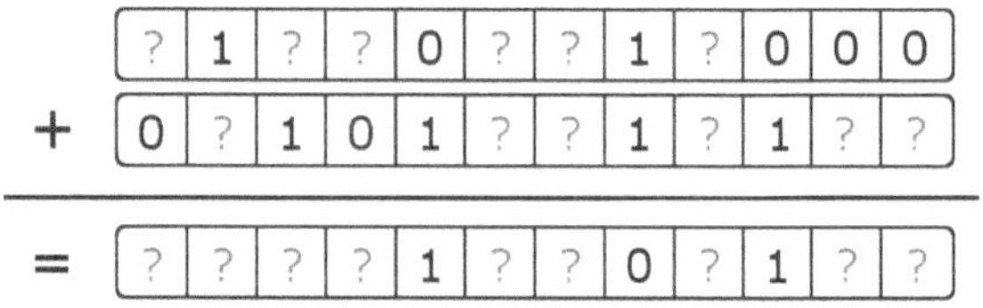

Fig. 8.3 Alice's information on the key is reduced by xor-ing several keys together

	?	1	?	?	0	?	?	1	?	0	0	0
+	0	?	1	0	1	?	?	1	?	1	?	?
=	?	?	?	?	1	?	?	0	?	1	?	?

Alice's measurement result	sifting set	guess of Bob's state	guess of Bob's bit
$\mid\uparrow\rangle$	$\{\mid\uparrow\rangle, \mid\rightarrow\rangle\}$	?	?
$\mid\uparrow\rangle$	$\{\mid\rightarrow\rangle, \mid\downarrow\rangle\}$	$\mid\rightarrow\rangle$	1
$\mid\uparrow\rangle$	$\{\mid\downarrow\rangle, \mid\leftarrow\rangle\}$	$\mid\leftarrow\rangle$	1
$\mid\uparrow\rangle$	$\{\mid\leftarrow\rangle, \mid\uparrow\rangle\}$	?	?

Information reduction: The bit string of length $k \times N$ is divided into k substrings of length N. These substrings are added bitwise, yielding a string of length N. The bitwise addition is commutative and acts as $+ \oplus + = - \oplus - = +, + \oplus - = -$, $+\oplus? = -\oplus? = ?$ (c.f. Fig. 8.3). If Alice is left with a string of question marks?, the protocol is restarted. If this happens too often, Bob aborts the protocol to avoid that Alice keeps only cases where she has few question marks.

Database access: Alice announces the number $s = j - i$ where j is the item of the database that she's interested in and i is an item of the xor-ed key K^f that she knows. Bob announces the N bits $C_i = X_i \oplus K^f_{i+s}$ where X_n are the elements of his database. Alice deduces the element she's interested in $X_i = C_i \oplus K^f_j$.

8.2.2 Discussion

As mention above, a private database query protocol must provide two kinds of security. First, the database holder needs some guarantees that little information about his full database is revealed during the protocol. To see why this is the case here, we realize that the only way for Alice to know elements of Bob's database is by guessing bits in the key K^f. But the states that Alice needs to discriminate for this, even after having learned the sifting sets, are not orthogonal to each other. An individual attack thus never allows her to learn Bob's bit with certainty. There is thus a bound on how much information on the database Alice has access to in this case. In [18] we discuss in more details how the reduction step ensures that the key hold by Alice contains many question marks, so that she cannot learn many elements of Bob's database.

Second, the user needs to make sure that the database holder has little chance of guessing the element of the database that she is interested in. In order to learn the item of the database that Alice is interested in, Bob needs to guess j, the item of Alice's final key that is different from a question mark. He thus needs to learn about the conclusiveness of Alice's transcoding. But the choice of Alice's measurement bases is unknown to Bob, he can thus never be sure whether she translated her result to a question mark or not. This remains partly true even in the case that Bob sends different states than the ones prescribed by the protocol (c.f. [18] for more details).

The above protocol for database queries thus provides some level of security for both the user and the database provider, while being resistant to losses. The exact amount of security provided is however not very clear yet. In particular, we only considered here specific individual attacks. It would thus be interesting on one side to study more general attacks, and on the other side to develop security proofs for given classes of attacks.

References

1. J. Mueller-Quade, R. Renner, New J. Phys. **11**, 085006 (2009)
2. C. Clausen, I. Usmani, F. Bussières, N. Sangouard, M. Afzelius, H. de Riedmatten, N. Gisin, Nature **469**, 508 (2011)
3. E. Saglamyurek, N. Sinclair, J. Jin, J.A. Slater, D. Oblak, F. Bussières, M. George, R. Ricken, W. Sohler, W. Tittel, Nature **469**, 512 (2011)
4. H.P. Specht, C. Nölleke, A. Reiserer, M. Uphoff, E. Figueroa, S. Ritter, G. Rempe, Nature **473**, 190 (2011)
5. B. Julsgaard, J. Sherson, J.I. Cirac, J. Fiurášek, E.S. Polzik, Nature **432**, 482 (2004)
6. N. Lütkenhaus, Phys. Rev. A **54**, 97 (1996)
7. A. Bocquet, R. Alléaume, A. Leverrier, J. Phys. A: Math. Theor. **45**, 025305 (2012)
8. D. Bruss, Phys. Rev. Lett. **81**, 3018 (1998)
9. U. Maurer, IEEE Trans. Inf. Theory **39**, 733 (1993)
10. I. Csiszár, J. Körner, IEEE Trans. Inf. Theory **24**, 339 (1978)
11. R. Ahlswede, I. Csiszár. IEEE Trans. Inf. Theory **39**, 1121 (1993)
12. H.-K. Lo, quant-ph/0102138
13. J. Kilian, in *Proceedings of 20th STOC*, ACM, New York, 1988, p. 20
14. B. Chor, O. Goldreich, E. Kushilevitz, M. Sudan, in *FOCS '95: Proceedings of the 36th Annual Symposium on Foundations of Computer Science*, 1995, p. 41
15. E. Kushilevitz, R. Ostrovsky, in *FOCS '97. Proceedings of the 38th Annual Symposium on Foundations of Computer Science*, 1997, p. 364
16. V. Giovannetti, S. Lloyd, L. Maccone, Phys. Rev. Lett. **100**, 230502 (2008)
17. V. Scarani, A. Acín, G. Ribordy, N. Gisin, Phys. Rev. Lett. **92**, 057901 (2004)
18. M. Jakobi, C. Simon, N. Gisin, C. Branciard, J.-D. Bancal, N. Walenta, H. Zbinden, Phys. Rev. A **83**, 022301 (2011)

Chapter 9
Finite-Speed Hidden Influences

The violation of a Bell inequality with space-like separated measurements precludes the explaination of nonlocal correlations in terms of causal influences propagating slower than light. Yet, these correlations can still be explained in a causal manner if one gives up Bell's locality condition. Indeed, this is the explanation followed when one says something like "A measurement on the singlet state $|\psi^-\rangle = \frac{1}{\sqrt{2}}(|01\rangle - |10\rangle)$ yielding result '0' in the computational basis of Alice prepares the state $|1\rangle$ for Bob". With a slightly different taste, Bohmian mechanics also provides a causal explanation for quantum correlations, which does not rely on quantum steering or collapse of the wavefunction. However, both of these explanations are much more nonlocal than a simple violation of Bell's local causality condition implies: not only do they involve faster-than-light influences at a distance, but these influences also have *immediate* effects on distant particles no matter how far away they are from each other. Here we question whether such a strong violation of the notion of locality is necessary or not.

9.1 Finite-Speed Propagation and v-Causal Theories

One way of violating Bell's local causality condition while still keeping a notion of "locality" is to allow causal influences to propagate faster than light, but only up to some finite speed $v < \infty$. In this way, instantaneous influence at a distance is avoided, and causal influences can still be understood as propagating in spacetime, i.e. acting locally, "de proche en proche".

Since the advent of special relativity, it might seem uncalled-for to consider faster-than-light propagations in space-time.[1] Indeed, it is well-known that faster-than-light information transmission in a Lorentz-invariant theory can generate temporal paradoxes [1]. However, this needs not be the case if the theory describing the interaction

[1] As a matter of fact, the same remark applies to instantaneous influences of the kind we just mentioned above.

J.-D. Bancal, *On the Device-Independent Approach to Quantum Physics*, Springer Theses,
DOI: 10.1007/978-3-319-01183-7_9,

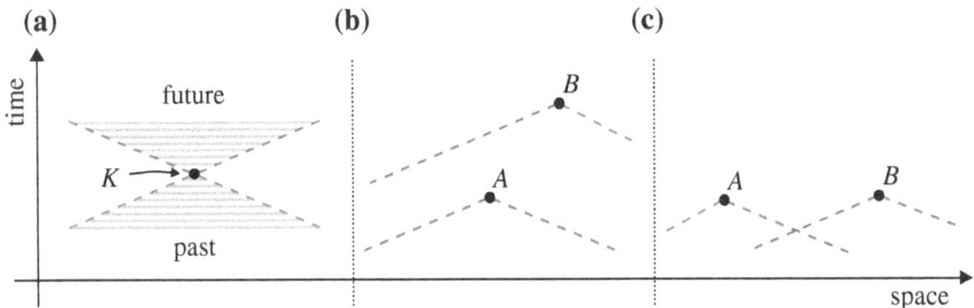

Fig. 9.1 Space-time diagram in the preferred reference frame. **a** The past and future v-cones (*hatched areas*) define the sets of events that can influence, or that can be influenced by K within a v-causal theory. **b** $A < B$: finite-speed influences can propagate from A to B. **c** $A \sim B$: no influence can be directly exchanged between two events which are not in each other's v-cones

with supraluminal transmissions is not Lorentz-invariant. For instance, if the speed of every faster-than-light communication is defined in a unique reference frame, then a temporal order is restored.

Considering thus a preferred reference frame for definiteness, we can formalize the idea of finite-speed causal influences as follows: a past and a future v-cone is associated to every event K in the preferred frame (c.f. Fig. 9.1a).What happens at K can only influence other events lying in the future v-cone of K, and K can only depend on what is contained within its past v-cone. We denote by $A < B$ configurations in which A lies in the past v-cone of B, and $A \sim B$ those in which A and B lie outside each-other's v-cones (c.f. Fig. 9.1b and c). Any theory satisfying these constraints is referred to as being v*-causal*. Note that Bell's condition of local causality is recovered for $v = c$.

9.1.1 v-Causal Models and Experimental Limitations

Clearly, v-causal theories, just like locally causal ones, are fundamentally incompatible with quantum physics. Indeed, they don't allow two parties to violate a Bell inequality if their measurements are performed simultaneously in the preferred frame, whereas quantum physics predicts that such inequalities can be violated independently of the space-time location of the measurements. Provided that correlations in nature agree with the quantum predictions, one could thus expect to be able to rule out v-causal models experimentally.

However this is not directly possible. Indeed, due to the finite accuracy inherent to every experimental manipulation, a v-causal model with sufficiently large speed v can always explain the experimental violation of a Bell inequality. Moreover, since quantum correlations are no-signalling, they can always be reproduced with the aid of the one-way communication available to v-causal models. Thus, quite on the contrary, if all correlations that can be observed in Bell-like experiments agree with the quantum predictions, then they can also be explained by a v-causal theory.

Experiments performed so far have thus only been able to put a lower bound on the speed v that is needed for the viability of v-causality. For instance, Salart et al. [2] and Cocciaro et al. [3] have shown that, if the speed of the earth in the preferred frame is less than $10^{-3}c$, then v-causal theories must have a speed v larger than 10,000 times the speed of light c.

Given that experimental results cannot rule out v-causal models directly, we examine below the potential physical consequences of these theories.

In the following we distinguish between several kinds of correlations. First, correlations are referred to as *easily accessible* in an experiment if they don't require very good synchronization between any measurements. All correlations that a v-causal model can freely choose because influences were able to propagate through all parties are of this kind. Second, *hardly accessible correlations* are those which require nearly perfect synchronization, the degree of synchronization required depending on the speed v of the model. Since some measurements are too simultaneous to allow influences to propagate between them in this case, v-causal models cannot produce all possible correlations of this kind. Finally, correlations are said to be *not directly accessible* if they require perfect synchronization between some measurements. In this case at least part of these correlations involve simultaneous measurements and are thus even *inaccessible* in principle.

A v-causal model is then said to be quantum if every time its correlations are easily accessible they are also in agreement with the quantum prediction. v-causal models which are not quantum can in principle be detected experimentally, whereas quantum v-causal models are experimentally indistinguishable from quantum physics without extraordinary synchronization capabilities.

Even though both easily accessible and hardly accessible correlations are in principle measurable, we would like to say something about v-causal models independently of their typical speed v, based only on the measurement of easily accessible correlations.

9.1.2 Influences Without Communication?

As mentioned in Chap. 1, the fact that faster-than-light influences be needed in order to reproduce some correlations does not necessarily allow these correlations to be used to signal faster than light. Rather, a violation of the no-signalling conditions (1.1.2) is needed to allow correlations to be used for communication. The superluminal influences of a v-causal model can thus remain hidden from observers having only access to the produced correlations if these correlations are no-signalling. In particular, as long as the correlations produced agree with the quantum predictions, they are no-signalling and thus cannot be used to communicate.

Since all easily accessible correlations produced by a quantum v-causal model are quantum, simultaneous measurements must be considered in order to allow quantum

v-causal models with arbitrary speed v to produce correlations diverging from the quantum prediction.

It was suggested in [4, 5] that the correlations predicted by a v-causal model in this situation could become signalling and allow for faster-than-light communication. Here, we investigate this question in more detail, and show that it is indeed possible to communicate faster than light in a v-causal world in which all easily accessible correlations are in agreement with quantum physics.

Note that a first example of situation in which a v-causal model was shown to allow for faster than light communication was put forward recently in [6]. However, this example requires the observation of supra-quantum correlations in order to conclude. It thus doesn't apply to quantum v-causal models, and is not likely to lead to an experimental application.

We present below a general approach which allows one to test if the existence of signalling correlations can be deduced from the knowledge of easily accessible correlations. We then examine whether such a test can be expected to be conclusive if correlations observed experimentally are assumed to be the ones predicted by quantum theory.

9.2 The Hidden Influence Polytope

Following the above discussion, we consider a space-time configuration in which some measurements are simultaneous in order to open the possibility for quantum v-causal models to produce non-quantum correlations. We then examine whether the correlations produced by the model in this configuration can remain no-signalling or not.[2]

For definiteness, let us consider here the 4-partite space-time configuration $R = (A < D < (B \sim C))$ shown in Fig. 9.2. A v-causal model in this situation must produce BC correlations that are local, even after conditioning on what happened at A and D (see Chap. 10 for more details). The correlations $P(bc|yz, axdw)$ must thus satisfy all bipartite Bell inequalities

$$\sum_{bcyz} \beta^i_{bcyz} P(bc|yz, axdw) \leq \beta^i_0 \tag{9.2.1}$$

where $\{(\beta^i_0, \beta^i_{bcyz})\}_i$ denote the coefficients of all Bell inequalities that are relevant given the number of inputs and outputs of each party.

[2] Note that signalling could be activated in cases where the model only produces no-signalling correlations as well. Indeed, if a marginal probability distribution can have different (no-signalling) values depending on the time chosen by some other party to perform his measurement, in the fashion of [4, 5], this change in the correlation can allow to guess the time of measurement chosen by a distant party. However we don't consider this possibility here.

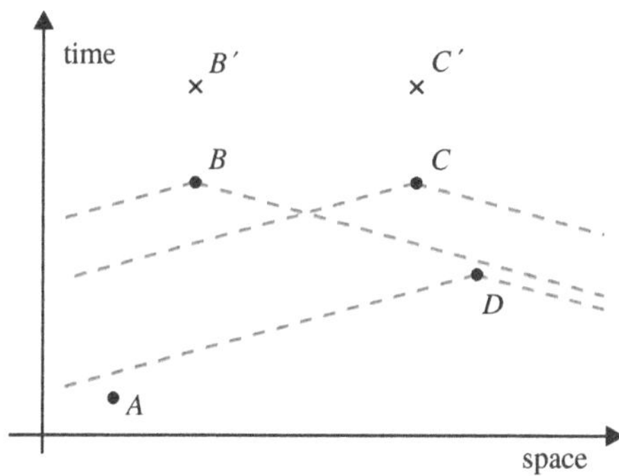

Fig. 9.2 In the four-partite Bell-type experiment characterized by the space-time ordering $R = (A < D < (B \sim C))$, no influence can be exchanged between Bob and Charly. However, if Charly delays his measurement, he can allow the configuration to recover a complete order $T_1 = (A < D < B < C')$. Similarly, Bob can delay his measurement in order to obtain the order $T_2 = (A < D < C < B')$

On the other hand, the correlations $P(abcd|xyzw)$ are no-signalling if and only if they satisfy the 4-partite no-signalling conditions:

$$\sum_a P(abcd|xyzw) = P(bcd|yzw) \;, \; \sum_b P(abcd|xyzw) = P(acd|xzw)$$
$$\sum_c P(abcd|xyzw) = P(abd|xyw) \;, \; \sum_d P(abcd|xyzw) = P(abc|xyz). \tag{9.2.2}$$

No-signalling correlations produced by a v-causal model in the R configuration must thus satisfy both condition (9.2.1) and (9.2.2). Since these form a finite set of linear conditions, they define a polytope in the space of correlations (c.f. Appendix A). We refer to this polytope as the *hidden influence polytope* associated to R.

In order to test whether a v-causal model satisfies the above conditions, we need to know which correlations the model produces in the R configuration. But since B and C are measured simultaneously in R, the correlations $P(abcd|xyzw)$ are not directly accessible: their observation requires perfect synchronization between some of the measurements. Still, given the properties of v-causal models, one can show that some parts of the 4-partite distribution $P(abcd|xyzw)$ can be inferred.

To see this, consider that Charly, in the experiment of Fig. 9.2, could perform his measurement at C as planed initially or choose to delay it to C' (or even to never do it). In any case, since he can in principle make his choice outside of the past v-cone of A, B and D, his choice cannot affect what happens at A, B and D. Thus, the ABD marginal produced by the model must be the same in the R configuration as in the $T_1 = (A < D < B < C')$ configuration. Since correlations in the T_1 configuration are easily accessible, the ABD marginal in the R configuration can be determined through measurements in T_1.

Similarly, one can show that the ACD marginal in the R configuration must equal that in the $T_2 = (A < D < C < B')$ configuration. It is thus also easily accessible, and both the ABD and the ACD marginal in the R configuration can in principle be known. The BC marginal is however clearly inaccessible experimentally since it explicitely requires measurements to be performed simultaneously. These two 3-partite marginals thus constitute the maximum amount of information that one can hope to infer about the R configuration.

In order to reach a conclusion without making assumptions on the value of the unknown marginals, we project the hidden influence polytope onto the subspace spanned by the ABD and ACD marginals. This allows us to deduce the conditions satisfied by the v-causal model in the situation R, in terms of the known marginals only.

Note that whenever an inequality satisfied by this projected polytope is violated, one of the two conditions (9.2.1) or (9.2.2) must be violated as well. Since (9.2.1) cannot be violated in the R configuration, by the definition of v-causality, the model must violate condition (9.2.2) in this configuration, i.e. produce signalling correlations.

Using techniques described in Appendix A, we could find several inequalities of the projected hidden influence polytope in configuration R when all parties use binary inputs and outputs. We present one of them below.

9.2.1 Quantum Violation and Faster-than-Light Communication

The following inequality is satisfied by all no-signalling correlations produced by a v-causal model in the R configuration (c.f. Fig. 9.2):

$$\begin{aligned} S = & - 3\langle A_0\rangle - \langle B_0\rangle - \langle B_1\rangle - \langle C_0\rangle - 3\langle D_0\rangle \\ & - \langle A_1B_0\rangle - \langle A_1B_1\rangle + \langle A_0C_0\rangle \\ & + 2\langle A_1C_0\rangle + \langle A_0D_0\rangle + \langle B_0D_1\rangle \\ & - \langle B_1D_1\rangle - \langle C_0D_0\rangle - 2\langle C_1D_1\rangle \\ & + \langle A_0B_0D_0\rangle + \langle A_0B_0D_1\rangle + \langle A_0B_1D_0\rangle \\ & - \langle A_0B_1D_1\rangle - \langle A_1B_0D_0\rangle - \langle A_1B_1D_0\rangle \\ & + \langle A_0C_0D_0\rangle + 2\langle A_1C_0D_0\rangle - 2\langle A_0C_1D_1\rangle \\ & \leq 7, \end{aligned} \tag{9.2.3}$$

where $\langle A_x\rangle = P_A(0|x) - P_A(1|x)$, $\langle A_xB_y\rangle = \sum_{ab}(-1)^{a+b}P_{AB}(ab|xy)$ and so on.

Recall that by construction this inequality only involves correlations that are easily accessible through some experiment. A quantum v-causal model can thus reproduce any value of S that is achievable with quantum correlations.

Interestingly, this inequality can be violated by measuring a 4-qubit state (c.f. Chap. 10). We can thus deduce that the corresponding quantum v-causal model must produce signalling correlations in the R configuration.

Thanks to the geometry of this configuration, any signalling obtained in the correlations can be used to communicate faster than light as soon as $v > c$. Indeed, by

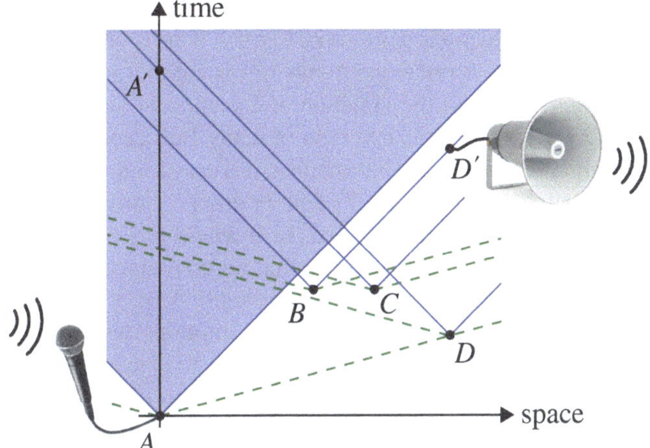

Fig. 9.3 In the configuration of Fig. 9.2, letting the parties B and C broadcast (at light-speed) their measurement results allows one to evaluate the marginal correlations BCD at the point D', which lies outside of the future light-cone of A (*shaded area*). If this marginal depends on Alice's input, it can be used for superluminal communication from A to D'. Similarly, if the ABC marginal correlations depend on the measurement w made at D, superluminal communication is possible from D to the point A'

definition of v-causal models, signalling can neither happen from B to ACD, nor from C to ABD, which lie in the past of C. It must thus happen either from A to BCD or from D to ABC. In both cases, this signalling in the correlations can be used to send signals faster than light (see Fig. 9.3).

Finite-speed v-causal models for quantum correlations thus allow for faster than light communication.

9.3 Experimental Perspectives

By construction, inequality (9.2.3) can be evaluated without requiring perfect synchronization between any of the four parties measured. It thus opens the possibility to test v-causal models experimentally in a way that is independent of the speed v, unlike precedent approaches. Namely, an experimental violation of Eq. (9.2.3) would allow one to conclude that if a v-causal model is responsible for the observed correlations, then it must also allow for faster than light communication in some situation.

Note that this conclusion is also valid if other systems than the four of interest happen to be measured during the experiment, even if this makes some measurements happen simultaneously in the privileged reference frame.

One way to evaluate the quantity S experimentally could be by performing measurements in the T_1 and T_2 configurations of Fig. 9.2. This is possible in principle if one knows how the preferred reference frame moves with respect to earth.

A simpler demonstration of the violation of S could also be performed without closing the locality loophole, as is common in many Bell experiments. Indeed, by performing measurements in a time-like manner, it is easy to ensure that the measurements are performed according to the T_1 and T_2 orders. However, this experiment would not be as strong as the previous ones as it would rely on a proper shielding of the measured systems in order to ensure that no communication between them happened by an exchange of some physical systems. Also, it would not allow to conclude directly (i.e. without invoking further assumptions) about the possibility to communicate faster-than-light. Rather, it would allow to conclude that slower-than-light signalling (communication without a physical support) is possible.

9.4 Conclusion

Thanks to inequality (9.2.3), we proved that the nonlocality of quantum correlations cannot be explained by superluminal finite-speed causal influences without opening the possibility to communicate faster than light. If one rejects this possibility, then one should also reject v-causal models as an attempt to keep a form of locality in causal explanations of quantum correlations.

Moreover, we argued that extraordinary synchronization between measurements is not necessary in order to reach this conclusion, independently of the speed v. This contrasts with previous approaches to v-causal models which could only test models with a speed v limited by technological constraints. It thus opens the possibility for new experimental approach to these models.

References

1. T. Mauldin, *Quantum Non-Locality and Relativity: metaphysical Intimations of Modern Physics* (Blackwell Publishers, Oxford, 2002)
2. D. Salart, A. Baas, C. Branciard, N. Gisin, H. Zbinden, Nature **454**, 861 (2008)
3. B. Cocciaro, S. Faetti, L. Fronzoni, Phys. Lett. A **375**, 379 (2011)
4. V. Scarani, N. Gisin, Phys. Lett. A **295**, 167 (2002)
5. V. Scarani, N. Gisin, Braz. J. Phys. **35**, 2A (2005)
6. S. Coretti, E. Hänggi, S. Wolf, Phys. Rev. Lett. **107**, 100402 (2011)

Chapter 10
Quantum Non-Locality Based on Finite-Speed Causal Influences Leads to Superluminal Signalling

The experimental violation of Bell inequalities using spacelike separated measurements precludes the explanation of quantum correlations through causal influences propagating at subluminal speed. Yet, any such experimental violation could always be explained in principle through models based on hidden influences propagating at a finite speed $v > c$, provided v is large enough. Here, we show that for *any* finite speed v with $c < v < \infty$, such models predict correlations that can be exploited for faster-than-light communication. This superluminal communication does not require access to any hidden physical quantities, but only the manipulation of measurement devices at the level of our present-day description of quantum experiments. Hence, assuming the impossibility of using nonlocal correlations for superluminal communication, we exclude any possible explanation of quantum correlations in terms of influences propagating at any finite speed. Our result uncovers a new aspect of the complex relationship between multipartite quantum nonlocality and the impossibility of signalling.[1]

Correlations cry out for explanation [1]. Our intuitive understanding of correlations between events relies on the concept of causal influences, either relating directly the events, such as the position of the moon causing the tides, or involving a past common cause, such as seeing a flash and hearing the thunder when a lightning strikes. Importantly, we expect the chain of causal relations to satisfy a principle of continuity, i.e., the idea that the physical carriers of causal influences propagate continuously through space at a finite speed. Given the theory of relativity, we expect moreover the speed of causal influences to be bounded by the speed of light. The correlations observed in certain quantum experiments call into question this viewpoint. [2]

---

[1] This chapter appeared as "J-D. Bancal et al., *Quantum non-locality based on finite-speed causal influences leads to superluminal signalling*, Nature Phys. **8**, 867 (2012)."

[2] "J.-D. Bancal, Y.-C. Liang, N. Gisin, *Group of Applied Physics, CH-1211 Geneva 4* (University of Geneva, Switzerland)". "S. Pironio, *Laboratoire d'Information Quantique* (Université Libre de Bruxelles, Belgium)". "A. Acin, *ICFO-Institut de Ciéncies Fotóniques, Castelldefels 08860 Castelldefels* (Barcelona, Spain); *ICREA-Institució Catalana de Recerca i Estudis Avançats, Barcelona, 08010*, (Barcelona, Spain)". "V. Scarani, *Centre for Quantum Technologies* (National University of Singapore, Singapore); *Department of Physics* (National University of Singapore, Singapore)".

J.-D. Bancal, *On the Device-Independent Approach to Quantum Physics*, Springer Theses,
DOI: 10.1007/978-3-319-01183-7_10,

When measurements are performed on two entangled quantum particles separated far apart from one another, such as in the experiment envisioned by Einstein, Podolsky, and Rosen (EPR) [2], the measurement results of one particle are found to be correlated to that of the other particle. Bell showed that if these correlated values were due to past common causes, then they would necessarily satisfy a series of inequalities [1]. But theory predicts and experiments confirm that these inequalities are violated [3], thus excluding any past common cause type of explanation. Moreover, since the measurement events can be spacelike separated [4–6], any influence-type explanation must involve superluminal influences [7], in contradiction with the intuitive notion of relativistic causality [8].

This nonlocal connection between distant particles represents a source of tension between quantum theory and relativity [8, 9], however, it does not put the two theories in direct conflict thanks to the *no-signalling* property of quantum correlations. This property guarantees that spatially separated observers in an EPR-type experiment cannot use their measurement choices and outcomes to communicate with one another. The complex relationship between quantum nonlocality and relativity has been the subject of intense scrutiny [7–10], but less attention has been paid to the fact that quantum nonlocality seems not only to invalidate the intuitive notion of relativistic causality, but more fundamentally the idea that correlations can be explained by causal influences propagating continuously in space. Indeed, according to the standard textbook description, quantum correlations between distant particles, and hence the violation of Bell inequalities, can in principle be achieved instantaneously and independently of the spatial separation between the particles. Any explanation of quantum correlations via hypothetical influences would therefore require that they "propagate" at speed $v = \infty$, i.e. "jump" instantaneously from one location to another as in real actions at a distance.

Is such an *infinite speed* a necessary ingredient to account for the correlations observed in Nature or could a finite speed v, recovering a principle of continuity, be sufficient? In particular, could an underlying theory with a limit v on the speed of causal influences reproduce correctly the quantum predictions, at least when distant quantum systems are within the range of finite-speed causal influences[11]? Obviously, any such theory would cease to violate Bell inequalities beyond some range determined by the finite speed v. At first, this hypothesis seems untestable. Indeed, provided that v is large enough, any model based on finite-speed (hidden) influences can always be made compatible with all experimental results observed so far. It thus seems like the best that one could hope for is to put lower-bounds on v by testing the violation of Bell inequalities with systems that are further apart and better synchronized [12, 13].

Here we show that there is a fundamental reason why influences propagating at a finite speed v may not account for the nonlocality of quantum theory: all such models give, for *any* $v > c$, predictions that can be used for faster-than-light communication. Importantly, our argument does not require the observation of non-local correlations between simultaneous or arbitrarily distant events and is thus amenable to experimental tests. Our results answer a long-standing question on the plausibility of finite-speed models first raised in [14, 15]. Progress on this problem was recently made in [16], where a conclusion with a similar flavor was obtained, but not for quantum theory. Technically, our approach is independent and different from the one in [16], which relies on "transitivity of nonlocality", a concept that has not yet found any application in quantum theory.

We derive our results assuming that the speed of causal influences v is defined with respect to a privileged reference frame (or a particular foliation of spacetime into spacelike hyperplanes). It should be stressed that whilst the assumption of a privileged frame is not in line with the spirit of relativity, there is also no empirical evidence implying its absence. In fact, even in a perfectly Lorentz-invariant theory, there can be natural preferred frames due to the non-Lorentz-invariant distribution of matter — a well-known example of this is the reference frame in which the cosmic microwave background radiation appears to be isotropic (see, eg., Ref. [17]). Moreover, note that there do exist physical theories that assume a privileged reference frame and are compatible with all observed data, such as Bohmian mechanics [18, 19], the collapse theory of Ghirardi, Rimini, and Weber [20] and its relativistic generalisation [21]. While both of these theories reproduce all tested (non-relativistic) quantum predictions, they violate the principle of continuity mentioned above (otherwise they would not be compatible with no-signalling as our result implies).

The models that we consider, which we call v-causal models, associate to each spacetime point K, a past and a future "v-cone" in the privileged frame, generalizing

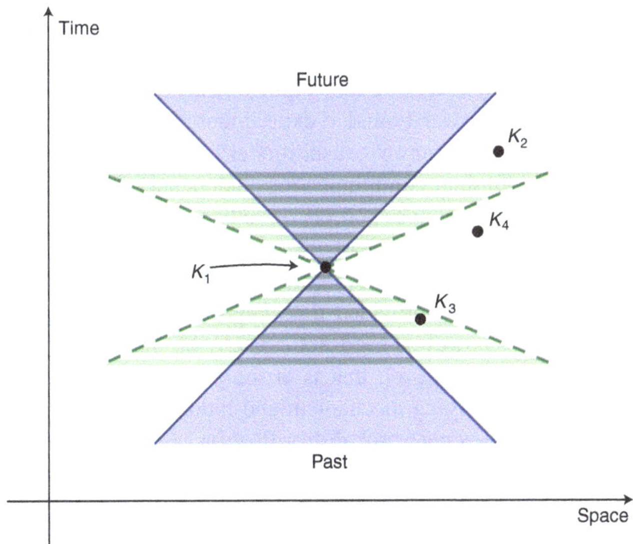

Fig. 10.1 Space-time diagram in the privileged reference frame. In the (shaded) light cone delimited by solid lines, causal influences propagate up to the speed of light c, whereas in the v-cone (hatched region), causal influences travel up to the speed v. An event K_1 can causally influence a spacelike separated event K_2 contained in its future v-cone and can be influenced by an event K_3 that lies in its past v-cone, but it cannot directly influence or be influenced by event K_4 outside its v-cone

the notion of past and future light-cones, see Fig. 10.1. An event at K_1 can have a causal influence on a point $K_2 > K_1$ located in its future v-cone and can be influenced by a point $K_3 < K_1$ in its past v-cone. But there cannot be any direct causal relation between two events $K_1 \sim K_4$ that are outside each other's v-cones. The causal structure that we consider here thus corresponds to Bell's notion of local causality [7, 22] but with the speed of light c replaced by the speed $v > c$. Operationally, it is useful to think of the correlations generated by v-causal models as those that can be obtained by classical observers using shared randomness together with communication at speed $v > c$.

According to the textbook description of quantum theory, local measurements on composite systems prepared in a given quantum state ρ yield the same joint probabilities regardless of the spacetime ordering of the measurements. However, a v-causal model will generally not be able to reproduce these quantum correlations when the spacetime ordering does not allow influences to be exchanged between certain pairs of events. In particular, the correlations between A and B will never violate Bell inequalities when $A \sim B$ (see Fig. 10.2). A possible programme to rule out v-causal models thus consists in experimentally observing Bell violations between pairs of measurement events as simultaneous as possible in the privileged reference frame [12]. As pointed out earlier, however, this programme can at best lower-bound the speed v of the causal influences.

More fundamentally, one could ask if it is even possible to conceive a v-causal model that reproduces the quantum correlations in the favourable situation where *all successive measurement events are causally related by v-speed signals*, that is, when any given measured system can freely influence all subsequent ones? In the bipartite case, this is always possible (see Fig. 10.2 and Chap. 10 in [23]), and thus the only possibility is to lower bound v experimentally. In the four-partite case, however, we show below that any v-causal model of this sort necessarily leads to the possibility of superluminal communication, independently of the (finite) value of v. Importantly, the argument does not rely directly on the observation of non-local correlations between simultaneous events.

Let us stress that v-causal models evidently allow for superluminal influences at the hidden, microscopic level, provided that they occur at most at speed v. Such superluminal influences, however, need not *a priori* be manifested in the form of signalling at the macroscopic level, that is at the level of the experimenters who have no access to the underlying mechanism and hidden variables λ of the model, but can only observe the average probability $P(ab|xy)$ (e.g., by rotating polarizers along different directions x, y and counting detector clicks a, b). It is this later sort of superluminal communication that we show to be an intrinsic feature of any v-causal model reproducing quantum correlations.

A sufficient condition for correlations P not to be exploitable for superluminal communication is that they satisfy a series of mathematical constraints known as the "no-signalling conditions". In the case of four parties (on which we will focus below), no-signalling is the condition that the marginal distributions for the joint system ABC are independent of the measurement performed on system D, i.e.,

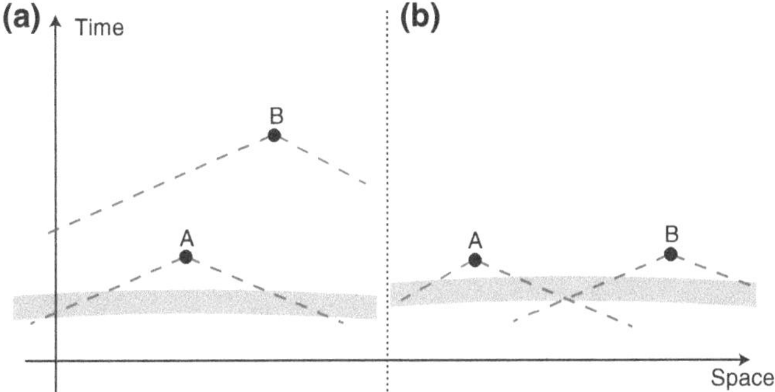

Fig. 10.2 Predictions of a v-causal model in a bipartite Bell experiment. We denote by $P(ab|xy)$ the probability associated to A and B observing respectively the outcomes a and b when their measurement is labeled by x and y. In quantum theory, such probabilities are given by $P_Q(ab|xy) = \mathrm{tr}(\rho M_a^x \otimes M_b^y)$, where ρ is the quantum state of A and B and M_a^x, M_b^y their respective measurement operators, and are independent of the space-time ordering of the measurements. In contrast, in a v-causal model, the observed probabilities will depend on the space-time ordering between A and B, as we now specify. a) A is in the past v-cone of B. Let the variable λ, with probability distribution $q(\lambda)$, denotes the joint state of the particles, or more generally a complete specification of any initial information in the shaded spacetime region that is relevant to make predictions about a and b (strictly, only the shaded region that is in the past v-cone of A can have a causal influence on A; however, all our arguments still follow through even if we consider spacetime regions of the kind depicted). In this situation we can write $P_{A<B}(ab|xy) = \sum_\lambda q(\lambda)P(ab|xy,\lambda) = \sum_\lambda q(\lambda)P(a|x,y\lambda)P(b|y,ax\lambda) = \sum_\lambda q(\lambda)P(a|x,\lambda)P(b|y,ax\lambda)$, where we used Bayes' rule in the second equality and the assumption that the measurement setting y is a free variable, i.e., uncorrelated to a, x, λ, in the last equality. Note that there always exists a trivial v-causal model that reproduces the quantum correlations in the case $A < B$ (or $B < A$) since we can write $P_Q(ab|xy) = P_Q(a|x)P_Q(b|y,ax)$ by the no-signalling property of quantum correlations (this easily generalises to the multipartite case, see Chap. 10 in [23]). b) A and B are outside each other's v-cones. As above, the variable λ represents a complete (as far as predictions about a and b are concerned) specification of the shaded spacetime region. Note that this region screens-off the intersection of the past v-cones of A and B, in the sense that given the specification of λ in the shaded region, specification of any other information in the past v-cones of A and B become redundant. It thus follows that $P(a|x,by\lambda) = P(a|x,\lambda)$ since any information about B is irrelevant to make predictions about a once λ is specified (see [7] for a more detailed discussion of this condition). Similarly $P(b|y,ax\lambda) = P(b|y,\lambda)$. We can therefore write $P_{A\sim B}(ab|xy) = \sum_\lambda q(\lambda)P(ab|xy,\lambda) = \sum_\lambda q(\lambda)P(a|x,y\lambda)P(b|y,ax\lambda) = \sum_\lambda q(\lambda)P(a|x,\lambda)P(b|y,\lambda)$. Formally, the correlations are thus "local" and satisfy all Bell inequalities. In particular, the model cannot reproduce arbitrary quantum correlations in this situation

$$\sum_d P(abcd|xyzw) = P(abc|xyz), \tag{10.1}$$

together with the analogous conditions for systems ABD, ACD, and BCD. Here $P(abcd|xyzw)$ is the probability that the four parties observe outcomes a, b, c and d when their respective measurements settings are x, y, z and w. These conditions imply that the marginal distribution for any subset of systems are independent of the measurements performed on the complementary subset.

Our main result is based on the following Lemma, whose proof can be found in Chap. 10 in [23].

Lemma *Let $P(abcd|xyzw)$ be a joint probability distribution with $a, b, c, d \in \{0, 1\}$ and $x, y, z, w \in \{0, 1\}$ satisfying the following two conditions.*

(a) The conditional bipartite correlations $BC|AD$ are local, i.e., the joint probabilities $P(bc|yz, axdw)$ for systems BC conditioned on the measurements settings and results of systems AD admit a decomposition of the form $P(bc|yz, axdw) = \sum_\lambda q(\lambda|axdw)P(b|y, \lambda)P(c|z, \lambda)$ for every a, x, d, w.

(b) P satisfies the no-signalling conditions (10.1).

Then there exist a four-partite Bell expression S (see Chap. 10 in [23] for its description) such that correlations satisfying (a) and (b) necessarily satisfy $S \leq 7$, while there exist local measurements on a four-partite entangled quantum state that yield $S \simeq 7.2 > 7$.

The Bell expression S has the additional property that it involves only the marginal correlations ABD and ACD, but does not contain correlation terms involving both B and C (this property is crucial for establishing our final result, as it implies that a violation of the Bell inequality can be verified without requiring the measurement on B and C to be simultaneous).

Consider now the prediction of a v-causal model in the thought experiment depicted in Fig. 10.3, where the space-time ordering between the parties in the privileged frame is such that $A < D < (B \sim C)$. Since B and C are outside each other's v-cones, it follows immediately that the $BC|AD$ correlations are local (see Chap. 10 in [23] for details). A violation of the Bell inequality $S \leq 7$ by the model in this configuration therefore implies that assumption (b) of Lemma 1 must be violated, i.e. that the correlations produced by the model violate the no-signalling conditions (10.1). It is easy to see that this further implies that these correlations can be exploited for superluminal communication (see caption of Fig. 10.3). It thus remains to be shown that the Bell inequality $S \leq 7$ is violated by a v-causal model in a configuration where $B \sim C$, as standard quantum theory suggests. Note that this should not be taken for granted since one should not a priori expect a v-causal model to reproduce the quantum correlations in such a situation, for the same reason that in the bipartite case we do not expect a v-causal model to reproduce the quantum correlations when $A \sim B$. Central to our argument lies the fact that the Bell expression S only involves the marginal correlations ABD and ACD, which allow ones, as we show below, to infer its value in a situation where $B \sim C$ from observations in which B and C are not necessarily measured outside each other's v-cones.

Explicitly, consider a modification of the thought experiment of Fig. 10.3, where the times t_B and t_C at which B and C are measured are chosen randomly so that any of the three configurations $A < D < B < C$, $A < D < C < B$, and $A < D < (B \sim C)$ can occur. Any v-causal model should at least reproduce the quantum correlations yielding $S \simeq 7.2 > 7$ in the first two situations, in which finite speed influences can freely travel from the first measured party to the last one. In particular, the v-causal model thus reproduces the marginal quantum correlations ABD when $A < D < B < C$. But then, it will also necessarily reproduce the same quantum

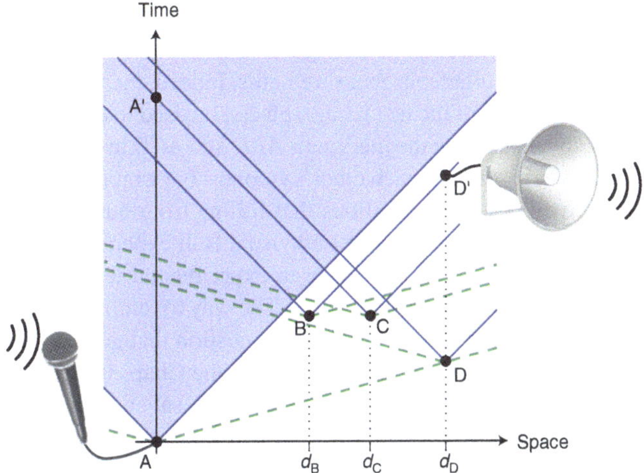

Fig. 10.3 Four-partite Bell-type experiment characterized by the spacetime ordering $R = (A < D < (B \sim C))$. Since B and C are both measured after A and D and satisfy $B \sim C$, the $BC|AD$ correlations produced by a v-causal model are local (see Chap. 10 in [23]). A violation of the inequality of Lemma 1 by the model therefore implies that the corresponding correlations must violate the no-signalling conditions (10.1). At least one of the tripartite correlations ABC, ABD, ACD, or BCD must then depend on the measurement setting of the remaining party. The marginal ABD (ACD) cannot depend on z (y), since this measurement setting is freely chosen at C (B), which is outside the past v-cone of A, B (C) and D (see also Chap. 10 in [23]). It thus follows that either the marginal ABC must depend on the measurement setting w of system D or that the marginal BCD must depend on the measurement setting x of system A (or both). Let the four systems lie along some spatial direction at, respectively, a distance $d_B = \frac{1}{4}(1 + \frac{1}{r}) + \frac{1}{1+r}$, $d_C = \frac{3}{4}(1 + \frac{1}{r}) - \frac{1}{1+r}$, $d_D = 1$ form A, where $r = v/c > 1$, and let them be measured at times $t_A = 0$, $t_B = t_C = \frac{2}{c+v}$, $t_D = 1/v$. Suppose that the BCD marginal correlations depend on the measurement x made on the first system A. If parties B and C broadcast (at light-speed) their measurement results, it will be possible to evaluate the marginal correlations BCD, at the point D'. Since this point lies outside the future light-cone of A (shaded area), this scheme can be used for superluminal communication from A to D'. Similarly, if the ABC marginal correlations depend on the measurement w made on D, they can be used for superluminal communication from D to the point A'

marginal in the situation $A < D < (B \sim C)$. Operationally, this is very intuitive: in both cases $B \sim C$ and $B < C$, the particle B can only use the shared randomness and the communication it received from A, D to produce its output. Furthermore, since it does not know when C is measured, it must produce the same output in both situations, hence the ABD marginal must be identical in both cases (see Chap. 10 in [23] for a more detailed argument). Similarly, we can infer that the quantum ACD marginal obtained for $A < D < C < B$ is reproduced when $B \sim C$. Together with the fact that the Bell expression S only involves the ABD and ACD marginals, a v-causal model must thus violate the inequality $S \leq 7$ in the configuration of Fig. 10.3, and hence give rise to correlations that can be exploited for superluminal communication.

In stark contrast with the bipartite scenario, these results therefore allow one to test experimentally the prediction of no-signalling v-causal models for any $v < \infty$ without requiring any simultaneous measurements. Indeed, the very same theoretical argument as that presented in the last paragraph can be used to deduce the value of S in the case $B \sim C$ by measuring the marginals ABD and ACD in situations in which B and C are not necessarily outside each other's v-cones. For a more detailed discussion on some of the experimental possibilities that follow from our result, we refer the reader to Chap. 10 in [23]. Note that as with usual Bell experiments, depending on the assumption that one is willing to take, an experimental test of v-causal model may also need to overcome various loopholes. The way to remove these assumptions and overcome these loopholes is an interesting question that goes beyond the scope of our work but some possibilities are discussed in the Chap. 10 in [23].

To conclude, we proved that if a v-causal model satisfies the requirement of reproducing the quantum correlations when the different systems are each *within* the range of causal influences of previously measured systems, then such a model will necessarily lead to superluminal signalling, for any finite value of $v > c$. Moreover, our result opens a whole new avenue of experimental possibilities for testing v-causal models. It also illustrates the difficulty to modify quantum physics while maintaining no-signalling. If we want to keep no-signalling, it shows that quantum nonlocality must necessarily relate discontinuously parts of the universe that are arbitrarily distant. This gives further weight to the idea that quantum correlations somehow arise from outside spacetime, in the sense that no story in space and time can describe how they occur.

Author Contributions All authors participated at various levels to the results presented here and to the writing of the article.

Author Information Reprints and permissions information is available at npg.nature.com/reprintsandpermissions. Correspondence and requests for materials should be addressed to J.-D. B. (jean-daniel.bancal@unige.ch)

Acknowledgments We acknowledge Serge Massar and Tamas Vértesi for helpful discussions as well as Jonathan Silman, Eric Cavalcanti, Tomy Barnea and Samuel Portmann for comments on the manuscript. This work was supported by the European ERC AG Qore and SG PERCENT, the European EU FP7 QCS and Q-Essence projects, the CHIST-ERA DIQIP project, the Swiss NCCRs QP & QSIT, the Interuniversity Attraction Poles Photonics@be Programme (Belgian Science Policy), the Brussels-Capital Region through a BB2B Grant, the Spanish FIS2010-14830 project, the National Research Foundation and the Ministry of Education of Singapore.

References

1. J.S. Bell, *Speakable and Unspeakable in Quantum Mechanics: Collected Papers on Quantum Philosophy* (Cambridge University Press, Cambridge, 2004)
2. A. Einstein, B. Podolsky, N. Rosen, Can quantum-mechanical description of physical reality be considered complete? Phys. Rev. **47**, 777–780 (1935)
3. A. Aspect, Bell's inequality test: more ideal than ever. Nature **398**, 189 (1999)
4. A. Aspect, J. Dalibard, G. Roger, Experimental test of bell's inequalities using time-varying analyzers. Phys. Rev. Lett. **49**, 1804–1807 (1982)
5. W. Tittel, J. Brendel, H. Zbinden, N. Gisin, Violation of bell inequalities by Photons more han 10 km apart. Phys. Rev. Lett. **81**, 3563–3566 (1998)
6. G. Weihs, T. Jennewein, C. Simon, H. Weinfurter, A. Zeilinger, Violation of bell's inequality under strict einstein locality conditions. Phys. Rev. Lett. **81**, 5039–5043 (1998)
7. T. Norsen, John. S. Bell's concept of local causality. Am. J. Phys. **79**, 1261–1275 (2012)
8. T. Mauldin, *Quantum Non-Locality and Relativity: Metaphysical Intimations of Modern Physics* (Blackwell Publishers, Oxford, 2002)
9. A. Shimony, *Search for a Naturalistic World View: Scientific Method and Epistemology* (Cambridge University Press, Cambridge, 1993)
10. S. Popescu, D. Rohrlich, Quantum nonlocality as an axiom. Found. Phys. **24**, 379–385 (1994)
11. D. Bohm, B.J. Hiley, *emphThe Undivided Universe* (Routledge, London and NY, 1993), p. 347
12. D. Salart, A. Baas, C. Branciard, N. Gisin, H. Zbinden, Testing the speed of 'spooky action at a distance'. Nature **454**, 861–864 (2008)
13. B. Cocciaro, S. Faetti, L. Fronzoni, A lower bound for the velocity of quantum communications in the preferred frame. Phys. Lett. A **375**, 379–384 (2011)
14. V. Scarani, N. Gisin, Superluminal influences, hidden variables, and signaling. Phys. Lett. A **295**, 167–174 (2002)
15. V. Scarani, N. Gisin, Superluminal hidden communication as the underlying mechanism for quantum correlations: Constraining models. Braz. J. Phys. **35**, 2A (2005)
16. S. Coretti, E. Hänggi, S. Wolf, Nonlocality is transitive. Phys. Rev. Lett. **107**, 100402 (2011)
17. C.H. Lineweaver et al., The dipole observed in the COBE DMR 4 year data. Astrophys. J. **470**, 38 (1996)
18. D. Bohm, A suggested interpretation of the quantum theory in terms of "hidden" variables. I Phys. Rev. **85**, 166–179 (1952)
19. D. Bohm, A suggested interpretation of the quantum theory in terms of "hidden" variables. II Phys. Rev. **85**, 180–193 (1952)
20. G.C. Ghirardi, A. Rimini, T. Weber, Unified dynamics for microscopic and macroscopic systems. Phys. Rev. D **34**, 470–491 (1986)
21. R. Tumulka, in *Quantum Mechanics: Are There Quantum Jumps? and On the Present Status of Quantum Mechanics* eds. by A. Bassi, D. Duerr, T. Weber, N. Zanghi. AIP Conference Proceedings 844 (American Institute of Physics, Melville, 2006), pp. 340–352
22. Bell, J. S. *Speakable and Unspeakable in Quantum Mechanics* Ch. 24 (Cambridge Univ. Press, 2004).
23. See supplementary information linked to the online version of the paper at www.nature.com/nphys.

Chapter 11
Outlook

In this thesis we presented several studies related to correlations in the context of quantum physics. First, we focused on general properties of correlations, the most important of which being the nonlocality of quantum correlations, i.e. the ability for results obtained upon measurement of a quantum system to violate a Bell inequality. While nonlocality has attracted a lot of attention in the bipartite case, our results indicate that the situation changes dramatically when a third party is considered.

Indeed, several of the results presented here don't have a bipartite analogue: it is impossible in the bipartite case to deduce that some global correlations are nonlocal by only studying their marginals (c.f. Sect. 3.3), no tight Bell inequality for two parties is known to be impossible to violate with quantum correlations (c.f. Sect. 3.5), and bipartite Bell experiments can only test v-causal models with a bounded speed v (c.f. Chaps. 9 and 10). This seems to indicate that much is still to be discovered in multipartite systems. For instance, it would be interesting to explore further constraints imposed by the relation between different marginals of a system.

Also, since several results obtained in the bipartite case don't extend straightforwardly to more parties, it could be interesting to look at these in more detail. Not much is known for instance about the possibility to simulate entangled states with classical resources in multipartite scenarios. Some physical principles like information causality, which generalizes the no-signalling principle to situations in which physical supports with bounded capacity are allowed to carry information, have also not yet found a good way to be expressed in multipartite scenarios [1–3]. Further investigation on these topics can give hints as to whether the difficulties encountered here are simply technical or whether they hide something more fundamental.

At a more technical level, given the important role played by polytopes in the characterization of correlations, the development of new tools to work with them would be very helpful. For instance, better ways to deal with symmetries of polytopes are highly desirable. Symmetries indeed typically induce a high level of redundancy in the description of polytopes, which makes several tasks on them highly inefficient.

In this thesis we also showed how working with correlations can provide robust conclusions in practical situations subject to uncertainties. The device-independent

DOI: 10.1007/978-3-319-01183-7_11,

assumptions are indeed weak enough to potentially allow for their implementation in practice, and yet strong enough to allow for the demonstration of interesting results, like the existence of genuinely multipartite entanglement. In other words, the ability to properly separate subsystems under study, and to identify their different possible inputs and outputs can be sufficient to obtain significant results.

However, experimental systems don't always meet these requirements. For instance, ions sharing the same trap can be hard to address individually, leading to an imperfect separation between subsystems (c.f. Sect. 6.3). Other systems, like superconducting qubits [4, 5], are subject to similar limitations. While this mismatch with the working assumptions could be seen as invalidating any possible conclusion, it also seems natural to expect small amounts of cross-talks between subsystems to have limited consequences. A proper way to estimate these cross-talks as well as a careful analysis of their possible impact would be welcome as it would allow one to easily apply the device-independent approach to many practical systems. However it remains to be found.

Finally, we also used correlations in this thesis as tools to study fundamental properties of nature. In particular, following J. S. Bell, we questioned the emergence of quantum nonlocal correlations in space-time. What our result suggests is that instead of asking how faster-than-light causal influences can coexist with the theory of relativity, we might just have to wonder about how infinitely-fast causal influences are at all compatible with relativity. Making the meaning of any of these questions more precise would already be a significative step forward.

References

1. M. Pawłowski, T. Paterek, D. Kaszlikowski, V. Scarani, A. Winter, M. Żukowski, Nature **461**, 1101–1104 (2009)
2. R. Gallego, L. E Würflinger, A. Acín, M. Navascués, Phys. Rev. Lett. **107**, 210403 (2011)
3. L.-Y. Hsu, Phys. Rev. A **85**, 032115 (2012)
4. M. Neeley, R.C. Bialczak, M. Lenander, E. Lucero, M. Mariantoni, A.D. O'Connell, D. Sank, H. Wang, M. Weides, J. Wenner, Y. Yin, T. Yamamoto, A.N. Cleland, J.M. Martinis, Nature **467**, 570 (2010)
5. L. DiCarlo, M.D. Reed, L. Sun, B.R. Johnson, J.M. Chow, J.M. Gambetta, L. Frunzio, S.M. Girvin, M.H. Devoret, R.J. Schoelkopf, Nature **467**, 574 (2010)

Appendix A
Polytopes

A.1 Definition and Terminology

A *polytope* $\mathcal{P} \subset \mathbb{R}^d$ is the convex hull of a finite number of points $V_i = (V_{i,1}, V_{i,2}, \ldots, V_{i,d}) \in \mathbb{R}^d$

$$\mathcal{P} = \{x = (x_1, x_2, \ldots, x_d) \in \mathbb{R}^d \text{ s.t. } \overline{x} = \sum_i q_i \overline{V_i},\ q_i \geq 0\} \tag{A.1.1}$$

where we denote by $\overline{x} = (1, x_1, x_2, \ldots, x_d)$ the points x completed by an extra component to fit in a space of dimension $d + 1$ for convenience [1].

In general, several sets of points $\{V_i\}_i$ can describe the same polytope $\mathcal{P}$ through (A.1.1). For instance, if a point V_i is not an extremal point of $\mathcal{P}$, i.e. if $\exists\ q_i \geq 0$ such that $\overline{V_j} = \sum_{i \neq j} q_i \overline{V_i}$ for some j, then $\{V_i\}_{i \neq j}$ describes the same polytope $\mathcal{P}$. On the other hand, if V_j is an extremal point of $\mathcal{P}$, then no set of points $\{V_i'\} \not\ni V_j$ can describe the same polytope $\mathcal{P}$. The description of a polytope through (A.1.1) is thus minimal when all point in $\{V_i\}_i$ are extremal points of the polytope, i.e. vertices. We refer to this as the *extremal points* description of a polytope, or *V-representation*. Notice that the condition for extremality of a point V_i is linear. The minimal set $\{V_i\}_{i,min}$ can thus be found from $\{V_i\}_i$ with the help of linear programming.

The *dimension* of a polytope $\dim(\mathcal{P})$ is given by the dimension of the smallest vector space that contains $\mathcal{P}$. It can be computed from the rank of its extremal points as

$$\dim(\mathcal{P}) = \begin{cases} \text{rk}(V_{ij}) & \text{if } \exists\ q_i \leq \mathbb{R} \text{ s.t. } \sum_i q_i \overline{V}_i = (1, 0, 0, \ldots, 0) \\ \text{rk}(V_{ij}) - 1 & \text{else.} \end{cases} \tag{A.1.2}$$

The main theorem on polytopes [1] tells that any polytope $\mathcal{P}$ can also be described as the intersection of finitely many *half-spaces* $\sum_j x_j H_{jk} \geq -H_{0,k}$, namely as:

J.-D. Bancal, *On the Device-Independent Approach to Quantum Physics*, Springer Theses,
DOI: 10.1007/978-3-319-01183-7,

$$\mathcal{P} = \{x \in \mathbb{R}^d \text{ s.t. } \sum_{j=0}^{d} \overline{x_j} H_{j,k} \geq 0 \ \forall k\} \tag{A.1.3}$$

As in the extremal point description of a polytope (A.1.1), the half-spaces description of a polytope can be made unique and minimal by requiring its inequalities to be *irredundant*, i.e. such that no $q_k \geq 0$ can satisfy $H_{j,k} = \sum_{k' \neq k} q_{k'} H_{j,k'}$.

When an inequality is irredundant, it is called a *facet* of the polytope. Its intersection with $\mathcal{P}$ is then of dimension $\dim(\mathcal{P}) - 1$. An inequality satisfied by the polytope which is not a facet might still have a non-null intersection with the polytope. The intersection of this inequality with the polytope is a often called a face of the polytope, and has a dimension strictly less than $\dim(\mathcal{P}) - 1$.

A polytope can thus be described in two equivalent ways (A.1.1), (A.1.3). Transforming one representation of a polytope into its dual one is in general a difficult task [2]. Nevertheless, when the polytopes are not too complicated, it can be possible to perform this transformation exactly with the aid of a computer. Several open-source softwares are available for this, like lrs [3], cdd [4], skeleton [5] or porta [6].

A.2 Some Operations on Polytopes

Polytopes can be manipulated in several ways. Here we describe some of these operations. See also the appendix of [7] for more examples.

A.2.1 Projection

One way to reduce the dimensionality of a polytope is to project it onto a subspace $S \subset \mathbb{R}^d$. For this, consider the linear projection operator $\Pi{:}\mathbb{R}^d \to S$. Without loss of generality, Π can be taken to act as $\Pi(x) = (x_1, x_2, \dots, x_s, 0, \dots, 0)$,[1] where $s = \dim(S)$ is the dimension of the projected space.

The projection of a polytope described in terms of extremal points is easily computed by projecting its extremal points:

$$\mathcal{P}' = \Pi(\mathcal{P}) = \left\{x \in \mathbb{R}^d \text{ s.t. } \overline{x} = \sum_i q_i \overline{\Pi(V_i)},\ q_i \geq 0\right\}. \tag{A.2.1}$$

Note that all projected vertices $\{\Pi(V_i)\}_i$ need not be extremal points of the projected polytope anymore. However all extremal points of $\mathcal{P}'$ are necessarily projections of some extremal points of the original polytope. They are thus necessarily contained

[1] A change of variables can be performed on order to let Π take this form if necessary.

in the set $\{\Pi(V_i)\}_i$. Projection of a polytope can thus only reduce the number of its extremal vertices.

When a polytope is specified in terms of half-spaces, finding the H-representation of its projection is more difficult. The Fourier-Motzkin algorithm achieves this without requiring to first transform the description of the polytope into its V-form, but it becomes quickly unpractical for larger problems because of its double-exponential computational complexity.[2]

Still, if one is not interested in the full set of inequalities describing the projected polytope $\mathcal{P}'$, a number of its facets can be found heuristically. Here we describe two linear programs that can be used for this.

The first linear program for finding facets of a projected polytope is similar to the shooting oracle described in [8]. The idea is, starting from a point that belongs to the interior of $\mathcal{P}'$, to travel as far as possible in one direction of the subspace S, until touching the boundary of $\mathcal{P}'$. The point then reached must generically belong to a facet of $\mathcal{P}'$.

Find a facet of a projected polytope. Let x^0 belong to the interior of $\mathcal{P}'$, and let $v \in S$ be a direction in the projected subspace. The linear program

$$\begin{aligned} \min_{z_k} & \sum_k H_{o,k} z_k + \sum_k \sum_{j=1}^{s} x_j^0 H_{j,k} z_k \\ \text{subject to} & \sum_{j=1}^{s} \sum_k v_j H_{j,k} z_k = -1 \\ & \sum_k H_{j,k} z_k = 0 \;\forall\; j = s+1, \ldots, d \\ & z_k \geq 0 \end{aligned} \tag{A.2.2}$$

provides the inequality

$$I_0 + \sum_{j=1}^{s} x_j I_j \geq 0 \tag{A.2.3}$$

with coefficients $I_0 = \sum_k H_{o,k} z_k + \sum_k \sum_{j=1}^{s} x_j^0 H_{j,k} z_k$, $I_j = \sum_k H_{j,k} z_k$ which is satisfied by all points x belonging to the projection of the polytope $\mathcal{P}'$. To see this one can check that the dual of this program is

[2] Note that other algorithms have been proposed, such as the ESP one [8] which is sensitive to the number of facets in the projected polytope $\mathcal{P}'$ rather than in the number of facets of the full polytope $\mathcal{P}$.

$$\max_{\mu, y_{s+1}, \ldots, y_d} \quad \mu$$
$$\text{subject to} H_{0,k} + \sum_{j=1}^{s} (x_j^0 + \mu v_j) H_{j,k} + \sum_{j=s+1}^{d} y_j H_{j,k} \geq 0 \; \forall \, k \tag{A.2.4}$$

which computes the largest value of μ such that $x^0 + \mu v \in \mathcal{P}'$.

In order to find a facet with the above linear program, one needs to choose a direction $v \in S$. Choosing an interesting direction is not necessarily obvious when nothing or just little about the projected polytope is known. Here is a linear program which can provide an interesting direction v to look for a facet when some facets of the projected polytope $\mathcal{P}'$ are known. The idea is that an interesting direction to look for a new facet of the projected polytope is in the direction of a vertex of the polytope described by the known facets of this polytope: if there is a difference between the known polytope and $\mathcal{P}'$, then some of its extremal points need to be outside $\mathcal{P}'$, otherwise the two polytopes are equal and there is nothing left to be found.

Find an extremal point of a polytope. Let $\{H_k\}_k$ be a set of inequalities defining a polytope $\mathcal{P}''$, and let $w \in \mathbb{R}^d$ be a direction in space. The following linear program yields a point x which lies on the boundary of $\mathcal{P}''$:

$$\min_{x_j} \quad \sum_j w_j x_j$$
$$\text{subject to} \quad H_{0k} + \sum_j H_{jk} x_j \geq 0 \; \forall \, k \tag{A.2.5}$$

If w is chosen at random, the point x is generically an extremal point of $\mathcal{P}''$.

Application to Polytope Representation Conversion

It is possible to find the H-representation of a polytope from its V-representation by performing a polytope slice (defined in the next section) followed by a polytope projection. Indeed, the conditions in Eq. (A.1.1) can be understood as defining a polytope $\tilde{\mathcal{P}} \in \mathbb{R}^{d+n}$, where n is the number of extremal points of $\mathcal{P}$, and $\tilde{\mathcal{P}}$ is defined by the following H-representation:

$$\tilde{\mathcal{P}} = \{(x, q) \in \mathbb{R}^{d+n} \text{ s.t.} \, \overline{x} = \sum_i q_i \overline{V}_i, \; q_i \geq 0\} \tag{A.2.6}$$

Elimination of the variables q_i, i.e. projection of $\tilde{\mathcal{P}}$ onto the subspace $S = \mathbb{R}^d$ thus defines the set of x such that the conditions in (A.1.1) hold. This projected polytope is thus $\mathcal{P}$, described in terms of inequalities. This is the basic technique used by the software porta [6] to solve a polytope.

A.2.2 Slice

Another way of reducing the dimensionality of a polytope is to consider a slice of it, that is, the intersection of the polytope $\mathcal{P}$ with a subspace $S \subset \mathbb{R}^d$ of dimension s. S can be defined by a set of linear equations: $S = \{x \in \mathbb{R}^d \text{ s.t. } \sum_j \overline{x}_j E_{jl} = 0 \,\forall l = 1 \ldots d - s\}$. The slice of $\mathcal{P}$ which belongs to S is easily defined if $\mathcal{P}$ is specified in the H-representation:

$$\mathcal{P}' = \left\{x \in \mathbb{R}^d \text{ s.t. } \sum_j \overline{x}_j M_{kj} \geq 0 \,\forall\, k, \sum_j \overline{x_j} E_{jl} = 0 \,\forall\, l\right\} \tag{A.2.7}$$

Computing the slice of a polytope in its V-representation is more complicated. In fact, on can show through polytope duality [1] that this task is equivalent to a projection of the polytope dual to $\mathcal{P}$ [9]. It can thus be done with the techniques described in the precedent section.

A.2.3 Another Task: Finding Facets Lying Under an Inequality

An inequality satisfied by a polytope is not a facet of the polytope if its dimension is lower than $d - 1$. When this is the case, one can show that the inequality can be expressed as the convex combination of a number of tighter inequalities, the tighter of which are facets of the polytope, sharing an intersection of dimension $d - 1$ with the polytope. Thus, if a point violates a non-facet inequality, it can only violate some of these tighter ones by a larger amount. Given a non-facet inequality, it can thus be interesting to look for these tighter facets.

A method for finding the facets underlying an inequality given a V-description of the polytope is presented in [10]. The idea is that the rank of the set of extremal points saturating the inequality can be augmented by adding more extremal points of the polytope to this set. When the achieved rank is sufficient, and if the hyperplane passing through these points does not cut the polytope into two parts, then it describes a facet of the polytope. By construction, the intersection of this facet with the polytope coincides with the intersection of the original inequality with the polytope since it is generated by the same set of extremal points.

Appendix B
Memoryless Attack on the 6-State Protocol: Proof

Here we provide a proof for the bound (8.1.4) used in the main text.

Proof Without loss of generality, Eve's POVM elements can be written as:

$$F_k = A_k^\dagger A_k = a_k \mathbb{1} + (b_k - a_k) P_k \tag{B.1}$$

where $A_k = \sqrt{a_k} U_k \overline{P_k} + \sqrt{b_k} U_k P_k$ is a Kraus operator associated to the element F_k, a_k, $b_k \geq 0$, U_k is a unitary operator, $\overline{P_k} = \mathbb{1} - P_k$ and P_k is a one-dimensional projector.

Eve's information on Alice's bit. We consider a given run k of the protocol for which Alice and Bob used the same basis $b = b_k$, and denote by A Alice's bit, B Bob's bit and E the result of Eve's POVM measurement. The total information gained by Eve on Alice's bit after the sifting procedure is given by $I(A : (E, b))$. Since b is independent of A and E, and since the state produced by Alice $\rho = \rho(A, b)$ is a function of A and b, we have:

$$\begin{aligned} I(A : (E, b)) &= H(A) + H(E, b) - H(A, E, b) \\ &= H(A) + H(E) + H(b) - H(A, E, b) \\ &= H(A, b) + H(E) - H(A, E, b) = I((A, b) : E) \\ &= I(\rho : E) = \log(6) - I(\rho|E) \end{aligned} \tag{B.2}$$

Where H is Shannon's entropy, and we assumed that the six states are chosen by Alice with the same probability $\frac{1}{6}$. We thus need to bound the quantity $I(\rho|E)$ which expresses the information that Eve is missing after she learns the result of her measurement, to know which state ρ was prepared by Alice. Using the fact that $\text{Prob}(E = k) = \text{tr}(F_k)/2$, this quantity can be expressed as

$$I(\rho|E) = I(E|\rho) + \sum_k \log(\text{tr}(F_k)) + \log(3) \tag{B.3}$$

J.-D. Bancal, *On the Device-Independent Approach to Quantum Physics*, Springer Theses, DOI: 10.1007/978-3-319-01183-7, © Springer International Publishing Switzerland 2014 where $I(E|\rho) = -\sum_{a,s,k} \text{Prob}(A = a, b = s, E = k) \log(\text{Prob}(E = k|A = a, b = s))$.

For every state ρ that Alice can produce, $\overline{\rho} = \mathbb{1} - \rho$ can also be produced by Alice. An attack described by the POVM elements $\{F_k\}$ thus provides Eve with the same information as the attack $\{\tilde{F}_k\}$ where $\tilde{F}_k = A_k \mathbb{1} + (b_k - a_k)\overline{P_k}$. Indeed, as shown above, this information is a symmetric function of $P(E = k|\rho = \rho(a, b))$ and

$$\begin{aligned} \text{tr}(\rho F_k) &= a_k + (b_k - a_k)\text{tr}(\rho P_k) \\ &= a_k + (b_k - a_k)\text{tr}(\overline{\rho}\tilde{F}_k) = \text{tr}(\overline{\rho}\tilde{F}_k). \end{aligned} \tag{B.4}$$

Moreover, the information that can be extracted from a mixture of measurements M_1 applied with probability p_1 and M_2 applied with probability p_2 is

$$I([(p_1, M_1), (p_2, M_2)]) = p_1 I(M_1) + p_2 I(M_2), \tag{B.5}$$

where $I(M_{1,2})$ is the information that can be extracted by using measurement $M_{1,2}$ only. We can thus write

$$\begin{aligned} I(\{F_k\}) &= I\left(\left\{\frac{1}{2}F_k\right\} \cup \left\{\frac{1}{2}\tilde{F}_k\right\}\right) \\ &= \sum_k \frac{a_k + b_k}{2} I\left[\left\{\frac{F_k}{a_k + b_k}, \frac{\tilde{F}_k}{a_k + b_k}\right\}\right] \end{aligned} \tag{B.6}$$

where the factor $1/(a_k + b_k)$ is a normalization coefficient. Thus, the information gained by performing an arbitrary POVM measurement can also be achieved by mixing measurement strategies consisting of only two POVM elements. Let us thus consider a POVM measurement for Eve consisting only of the two elements $\frac{F_k}{a_k+b_k}$ and $\frac{\tilde{F}_k}{a_k+b_k}$.

By direct computation one finds that

$$I(E_k|\rho) = \frac{1}{3}(h(c_k') + h(d_k') + h(e_k')) \tag{B.7}$$

with $c_k' = \frac{a_k+\text{tr}(\rho_2 P_k)(b_k-a_k)}{a_k+b_k}$, $d_k' = \frac{a_k+\text{tr}(\rho_4 P_k)(b_k-a_k)}{a_k+b_k}$, $e_k' = \frac{a_k+\text{tr}(\rho_6 P_k)(b_k-a_k)}{a_k+b_k}$. Since this function is convex in c', d', e', its minimum lies on the boundary of the admissible region

$$\left(\text{tr}(\rho_2 P_k) - \frac{1}{2}\right)^2 + \left(\text{tr}(\rho_2 P_k) - \frac{1}{2}\right)^2 + \left(\text{tr}(\rho_2 P_k) - \frac{1}{2}\right)^2 \leq \frac{1}{4}. \tag{B.8}$$

More precisely, this is found for $\text{tr}(\rho_2 P_k) \in \{0, 1\}$, $\text{tr}(\rho_4 P_k) \in \{0, 1\}$, or $\text{tr}(\rho_6 P_k) \in \{0, 1\}$. In this case, since $\log\left(\text{tr}\left(\frac{F_k}{a_k+b_k}\right)\right) = 0$, we find

$$I(\rho|E_k) = \frac{2 + h\left(\frac{b_k}{a_k+b_k}\right)}{3} + \log(3) \tag{B.9}$$

where $\epsilon_k = \frac{a_k}{b_k}$.

All in all, this gives the following bound on Eve's information about Alice's bit:

$$I(A:(E,b)) \leq 1 - \sum_k \frac{a_k + b_k}{2} \cdot \frac{2 + h\left(\frac{1}{1+\epsilon_k}\right)}{3}. \tag{B.10}$$

Perturbation on Bob's system. The attack of Eve delivers the state $\rho_i' = \sum_k A_k \rho_i A_k^\dagger$ to Bob instead of the expected ρ_i. This creates some errors in the outcomes of Bob, which are measured by the QBER:

$$Q = 1 - \sum_{i=1}^{6} P(A = i, B = i) = 1 - \frac{1}{6}\sum_i \mathrm{tr}(\rho_i' E_i) \tag{B.11}$$

where $E_i, i = 1, \ldots, 6$ are the six possible measurement operators of Bob and

$$\rho_i' = \sum_k \sqrt{a_k b_k} U_k \rho_i U_k^\dagger + \sqrt{a_k}(\sqrt{a_k} - \sqrt{b_k}) U_k \overline{P}_k \rho_i \overline{P}_k U_k^\dagger + \sqrt{b_k}(\sqrt{b_k} - \sqrt{a_k}) U_k P_k \rho_i P_k U_k^\dagger. \tag{B.12}$$

Note, that the attack $\{A_k\}$ has the same effect as $\{\tilde{A}_k\}$ for $\tilde{A}_k = \sqrt{a_k} U_k P_k + \sqrt{b_k} U_k \overline{P}_k$. Indeed, for all i there is a j such that $\overline{\rho}_i = \rho_j$ and $\overline{E}_i = E_j$, and one can check that:

$$\begin{aligned} \mathrm{tr}(U_k \rho_i U_k^\dagger E_i) &= \mathrm{tr}(U_k \overline{\rho}_i U_k^\dagger \overline{E}_i), \\ \mathrm{tr}(U_k P_k \rho_i P_k U_k^\dagger E_i) &= \mathrm{tr}(U_k \overline{P}_k \overline{\rho}_i \overline{P}_k U_k^\dagger \overline{E}_i). \end{aligned} \tag{B.13}$$

So we can assume that both A_k and $\tilde{A}_k$ are present in the attack. In this case the perturbed state is

$$\rho_i' = \sum_k \sqrt{a_k b_k} U_k \rho_i U_k^\dagger + \frac{(\sqrt{a_k} - \sqrt{b_k})^2}{2} (U_k \overline{P}_k \rho_i \overline{P}_k U_k^\dagger + U_k P_k \rho_i P_k U_k^\dagger). \tag{B.14}$$

To bound the second part of this expression, one can show by direct computation that

$$\sum_i \mathrm{tr}(U_k \overline{P}_k \rho_i \overline{P}_k U_k^\dagger E_i) + \mathrm{tr}(U_k P_k \rho_i P_k U_k^\dagger E_i)$$
$$= \sum_i \mathrm{tr}(\overline{P}_k \rho_i)\mathrm{tr}(\overline{P}_k U_k^\dagger E_i U_k) + \mathrm{tr}(P_k \rho_i)\mathrm{tr}(P_k U_k^\dagger E_i U_k) \tag{B.15}$$
$$= 2((1-c)(1-\tilde{c}) + c\tilde{c} + (1-d)(1-\tilde{d}) + d\tilde{d} + (1-e)(1-\tilde{e}) + e\tilde{e})$$

where $c = \mathrm{tr}(P_k\rho_2)$, $d = \mathrm{tr}(P_k\rho_4)$, $e = \mathrm{tr}(P_k\rho_6)$, $\tilde{c} = \mathrm{tr}(U_k P_k U_k^\dagger E_2)$, $\tilde{d} = \mathrm{tr}(U_k P_k U_k^\dagger E_4)$ and $\tilde{e} = \mathrm{tr}(U_k P_k U_k^\dagger E_6)$. The maximum value of (B.15) under the constraints $\left(c - \frac{1}{2}\right)^2 + \left(d - \frac{1}{2}\right)^2 + \left(e - \frac{1}{2}\right)^2 \leq \frac{1}{4}$, $\left(\tilde{c} - \frac{1}{2}\right)^2 + \left(\tilde{d} - \frac{1}{2}\right)^2 + \left(\tilde{e} - \frac{1}{2}\right)^2 \leq \frac{1}{4}$ can be checked to be 4.

Finally, the first part of (B.14) can also be bounded since $\mathrm{tr}(U_k\rho_i U_k^\dagger E_i) \leq 1$. Thus we find that

$$Q \geq \sum_k \frac{a_k + b_k}{2} \cdot \frac{(1-\sqrt{\epsilon_k})^2}{1+\epsilon_k}. \tag{B.16}$$

Putting the two bounds together. Let us consider equations (B.10) and (B.16) together. Keeping the sum $a_k + b_k$ constant for all k, we choose two values of k if possible: $k1$ and $k2$ such that $\epsilon_{k1} < \epsilon_{k2}$. Following [11] one can show that increasing ϵ_{k1} in such a way that keeps the bound on the QBER (B.16) unchanged, can only decrease ϵ_{k2} and increase Eve's information as given by (B.10). It is thus always better to have $\epsilon_k = \epsilon \; \forall \, k$. In this case both bounds become:

$$I(A:(E,S)) \leq \frac{1 - h\left(\frac{1}{1+\epsilon}\right)}{3}, \quad Q \geq \frac{(1-\sqrt{\epsilon})^2}{3(1+\epsilon)} \tag{B.17}$$

which can be summarized as

$$I(A:(E,S)) \leq \frac{1}{3}\left[1 - h\left(\frac{1-\sqrt{3Q(2-3Q)}}{2}\right)\right]. \tag{B.18}$$

Tightness of the bound. To show that the above bound is tight, consider the attack in which Eve uses the two POVM elements $F_k = \frac{1-\gamma}{2}\mathbb{1} + \gamma|k\rangle\langle k|$ for $k = 0, 1$. This gives $P(\rho = \pm z|k) = (1 \pm (-1)^k\gamma)/6$, $P(\rho = \pm x|k) = P(\rho = \pm y|k) = 1/6$, and so $I(A|E) = \frac{2}{3} + \log 3 + \frac{1}{3}h\left(\frac{1-\gamma}{3}\right)$. Since $I(A) = 1 + \log 3$, this attack proved Eve with a mutual information with Alice of $I(A:E) = I(A) - I(A|E) = \frac{1}{3}\left[1 - h\left(\frac{1-\gamma}{2}\right)\right]$. Moreover, the QBER induced by this attack is $Q = \frac{1-\sqrt{1-\gamma^2}}{3}$. Thus Eve can choose an attack that saturates Eq. (8.1.4): the bound is tight.

References

1. G.M. Ziegler, *Lectures on Polytopes, Graduate Texts in Mathematics 152*, Revised edn. (Springer, Berlin, 1998)
2. D. Bremner, On the complexity of vertex and facet enumeration for convex polytopes, PhD thesis (1998)
3. http://cgm.cs.mcgill.ca/~avis/C/lrs.html
4. http://www.ifor.math.ethz.ch/~fukuda/cdd_home/index.html
5. http://www.uic.unn.ru/~zny/skeleton/
6. http://typo.zib.de/opt-long_projects/Software/Porta/
7. T. Fritz, Polyhedral duality in Bell scenarios with two binary observables. J. Math. Phys. **53**, 072202 (2012) arXiv:1202.0141
8. http://www-control.eng.cam.ac.uk/~cnj22/research/projection.html
9. N.Yu. Zolotykh, private communication
10. J.-D. Bancal, N. Gisin, S. Pironio, J. Phys. A: Math. Theor. **43**, 385303 (2010)
11. N. Lütkenhaus, Phys. Rev. A **54**, 97 (1996)

Zeitfracht Medien GmbH
Ferdinand-Jühlke-Straße 7
99095 Erfurt, Deutschland
produktsicherheit@kolibri360.de